Unheeded Warning

The Inside Story of American Eagle Flight 4184

Stephen A. Fredrick
American Eagle ATR Pilot

McGraw-Hill

New York San Francisco Washington, D.C. Auckland Bogotá
Caracas Lisbon London Madrid Mexico City Milan
Montreal New Delhi San Juan Singapore
Sydney Tokyo Toronto

Mcgraw-Hill

A Division of The McGraw-Hill Companies

©1996 by **The McGraw-Hill Companies, Inc.**

Printed in the United States of America. All rights reserved. The publisher takes no responsibility for the use of any materials or methods described in this book, nor for the products thereof.

hc   1 2 3 4 5 6 7 8 9 FGR/FGR 9 0 0 9 8 7 6

Library of Congress Cataloging-in-Publication Data
ISBN 0-07-021951-6

McGraw-Hill books are available at special quantity discounts to use as premiums and sales promotions, or for use in corporate training programs. For more information, please write to the Director of Special Sales, McGraw-Hill, 11 West 19th Street, New York, NY 10011. Or contact your local bookstore.

Acquisitions editor: Shelley IC. Chevalier
Editorial team: Susan W. Kagey, Editor
      Lori Flaherty, Executive Editor
      Joann Woy, Indexer
Production team: Katherine G. Brown, Director
      Rose McFarland, Desktop Operator
      Nancy K. Mickley, Proofreading
Design team: Jaclyn J. Boone, Designer
      Katherine Lukaszewicz, Associate Designer

0219516
GEN1

To my fallen friends, Captain Orlando Aguiar, First Officer Jeffrey Gagliano, Flight Attendant Sandi Modaff, Flight Attendant Amanda Holberg, and the 64 passengers of American Eagle Flight 4184—"A Flight To Eternity."

A special thanks to my family for their encouragement, particularly my loving wife, Linde, and my mother, Shirley Quinn, for their understanding, inspiration, and support as my life took a dramatic turn, living the events that shaped and created this book.

Contents

Acknowledgments

I'd like to gratefully acknowledge the generous contributions to this work by the many individuals who only wished for the righting of what we all see as a terrible wrong. These people, and many more unnamed, did not seek recognition or notoriety, but they have earned my respect and thanks.

To Gary Washburn of the *Chicago Tribune,* for his forthright, honorable handling of the early story and continued caring. To Erin Hayes of ABC News, Chicago, for her genuine concern and integrity. To Steve Engelberg and Adam Bryant of the *New York Times,* for their tenacious pursuit of the truth. To William Maready, a man of ideals and conviction, whom I respect and admire, and whose respect I value most of all. To my sister, Cheri Kleffner, for her diligent work on the art in this book.

Author's note

The author has used participant recollections of conversations and quotations when available. Operational activities are based on reports or the author's personal experience as an airline pilot flying ATR aircraft for American Eagle based at Chicago O'Hare. Any conjecture about what would have been happening at a particular moment is noted in the text.

Conversations among crew members are based on personal experience as well as transcripts from cockpit voice recorders. The reconstruction of Flight 4184 is based on information provided by the flight data recorder, the cockpit voice recorder, the Federal Aviation Administration's transcript of tapes from the air traffic control facilities, including Clearance Delivery, Ground Control, Indianapolis Center and Chicago Center, and the National Transportation Safety Board's video reconstruction of the flight, presented at the initial hearing in Indianapolis in February 1995.

Foreword

With few exceptions, four o'clock in the afternoon of October 31, 1994, passed without notice to most people. In Montreal, a group of friends began gathering for a Halloween celebration, complete with costumes for the occasion. In a suburb of Chicago, children gleefully prepared for an unexpected trick-or-treating excursion with their father—unexpected because their daddy had been away on business and was not originally scheduled to be back in time.

In another house in another Chicago suburb, Halloween candy lay on the table, ready for distribution to neighborhood children. The young woman who bought the candy had agreed to travel on business only with the understanding that she would return in time to greet the trick-or-treaters.

In Dallas, a husband awaited the late-evening return of his wife and looked forward to hearing her pride-filled stories of her son, the new attorney, and her weekend with him and his family. The recently ill mother of another participant in the day's swearing-in ceremony anticipated congratulating her son when he and his father returned to Chicago.

In Houston, a mother thought about her daughter on her first trip as a flight attendant, and in Indianapolis, another mother sat and sewed her daughter's wedding dress for the scheduled event the following month, remembering the joys of the most recent visit of the soon-to-be married couple.

Somewhere in eastern North Carolina, the wonderful aroma of a roasting turkey filled a home in which the meal was timed to be ready when the traveling executive returned home to his wife and younger of two sons.

In a rental car halfway between Indianapolis and Chicago, a young woman and her son traveled northward through the wind-driven, icy rain.

In a small Korean village, an extended family greeted the dawn of the new day. In Scotland, a brother prepared to sleep, while farther south on the largest British Isle, a son concluded his day with a thought of his father.

In a Holiday Inn outside Cedar Rapids, Iowa, an airline pilot napped in anticipation of an early wake-up call the next morning. Near the Dallas-Fort Worth Airport, a young airline dispatcher sat at a computer terminal in a small cubicle that defined his workspace and monitored the progress of several flights while he prepared for the release of the next block. In Aurora, Illinois, at the Chicago Air Traffic Control Radar Facility, a controller calmly called on her radio for Eagle Flight 4184.

At four o'clock, October 31, 1994, these people were united by a tragedy that touched all their lives. This is their story.

1

Blind trust

At 3:57 P.M. on Halloween evening, October 31, 1994, American Eagle Flight 4184, an ATR-72 turboprop airliner carrying 64 passengers and a crew of four, crashed into a soybean field south of Roselawn, Indiana. The crash killed all 68 people on board.

Several witnesses caught a glimpse of the aircraft as it plunged out of a low overcast sky, and the sound of the impact could be heard for miles. A deer hunter in a tree stand described what he heard as a clap of thunder followed by the sound of a whining engine and then a plop like a piece of plywood being dropped onto the wet ground. On a farm more than a mile from the crash site, a woman reported that the windows in the house rattled as she helped her children get ready for Halloween. She thought at the time that it was thunder.

From under the hood of an old farm truck, hired hand Robert Hilton heard the screaming noise of what he thought was a small plane flying too low. Frightened by the peculiar sound so close, he jumped back from his work and scanned the dreary sky. Just to the east of the farm workshop, he spotted for an instant an aircraft at treetop level. In horror, he watched a scene that continues to haunt him. He was the only person to see Eagle Flight 4184 slam into the earth.

The air traffic controller monitoring traffic around Chicago O'Hare Airport was caught off guard as the electronic blip of Flight 4184 disappeared from her scope. She repeatedly called to the aircraft, and discussions with other controllers and other aircraft confirmed her fears—Flight 4184 had disappeared from radar coverage.

Local emergency crews arrived on site within minutes of the crash and were shocked by the wreckage left by a large aircraft that obviously contained no survivors. Television carried the story of a downed airliner on the 5 o'clock news, and almost immediately the media began to speculate as to the cause of the tragedy. As soon as the aircraft was confirmed to be a commuter airliner, reporters began talking about inexperienced crews and bad safety records in comparison to major airlines.

At the time of the accident, I was an American Eagle ATR first officer based at Chicago O'Hare Airport. At 5:15 P.M. on October 31, 1994, I was into my third night of a four-day trip and was dozing in my motel room in Cedar Rapids, Iowa, when the phone rang. I sleepily answered and heard the relieved sigh of my mother. She relayed the tragic news that my airline had lost an aircraft one hour earlier. I said a short, silent prayer and hoped for a mistake. I frantically searched for the television remote control and tuned to CNN. The press was speculating about a possible inflight failure of the aircraft because of the heavy weather. I had a bad feeling that I knew differently.

On the evening of November 23, 1993, the Tuesday before Thanksgiving, I was sitting in the right seat of an ATR-72 parked at Chicago O'Hare, awaiting our departure as American Eagle Flight 4165 to Marquette, Michigan, with a stop at Traverse City. I was teamed with an experienced captain, Dan Rodts, and flight attendants Susan Larson and Julie Bonk. We were heavy for the flight to Traverse City, and Dan and I reviewed the weather for the flight as the passengers boarded. Traverse City was blanketed by broken clouds, and we would make most of the trip above the overcast in a star-filled night sky.

This flight was one of American Eagle's stand-up trips that required us to report at 7:20 P.M. for an 8:20 P.M. departure. We would arrive in Marquette at midnight, get to the motel for four or five hours of rest, and take the early morning flight out of Marquette back through Traverse City to Chicago O'Hare, arriving at 8:00 A.M. Often, we were empty between Traverse City and Marquette, which gave the flight attendants some rest, although the pilots had to be awake. The stand-up trips met the letter of the law, but they were fatiguing and hotly debated by our pilot group. This particular trip was called "the stand-up from hell." This night's flight was our second of three in a row that scheduled us to be home for Thanksgiving by 8:00 in the morning.

Our flight to Traverse City was uneventful, and Dan allowed the autopilot to handle the aircraft while we discussed various subjects. I expressed my hope that Traverse would allow us to depart ahead of schedule so we could get some extra sleep in Marquette. Dan concurred.

The air was cold and smooth, and the northern lights were visible on the horizon. Our ATR-72 fought a headwind as we droned to the northeast over Lake Michigan, but we progressed quickly in the darkness. As we neared the eastern shore of the lake, I obtained the current weather information for Traverse City and called our station to let them know we were in range, requesting an expedited turn for our departure to Marquette. However, five passengers were to board, and they had not yet arrived at the airport.

The weather allowed us to use a visual approach into the airport at Traverse City, and Dan controlled the autopilot for the descent. We hoped to avoid penetrating much of the cloud deck to save a time-consuming deicing during our stop. "We've got passengers going to Marquette," I told Dan. "Who would book a flight from Traverse to Marquette tonight?"

"You mean they are boarding in Traverse?" he asked.

"Yeah," I said. "They just had better not be nonrevenue customers." I reached up to press the flight attendant call button and a chime sounded in the cabin.

"Hi," Julie said moments later. "Are we in range?"

"Yep. We also have five going to Marquette tonight, so you can't sleep on the next leg like you usually do," I chided good-naturedly.

"But all our passengers are getting off in Traverse," Julie said. "We checked."

"They have five reservations from Traverse to Marquette. Folks haven't checked in yet. Seat belts coming on. See you on the ground."

With the flight attendants notified, we ran through our checklists, and Dan briefed me for a VOR approach to Traverse City with a landing on the long runway, runway 28, heading west. We approached from the south, passed over the VOR and descended through the cloud deck until we saw the airport, then circled to the east. We landed at Traverse 10 minutes later, according to plan, and parked at the gate area. One of the ramp agents connected the ground power unit while the other handed our weather packet, dispatch release, and manifest to the captain through the small document door on the left side of the fuselage. Dan shut down both engines, and we heard the familiar thud of the cabin door being lowered to the frozen asphalt.

As we reviewed the weather package, it was obvious Marquette weather was as forecast, with a low ceiling at 600 feet above the ground, light snow, and a reported 3 miles of visibility. Temperature was 25 degrees Fahrenheit, and the wind was out of the northeast at 10 miles per hour. All in all, not a bad late-fall night in Marquette. We completed our paperwork, and Dan asked whether I could see any ice on the airplane or the icing evidence probe, which was located over my right shoulder. "Nothing visible, and I saw nothing accreting as we approached," I reported. "I'll go outside and look, just to be sure."

"You don't have to do that."

"No problem. We have the time, and I could use a breath of fresh air. I've been in this cockpit for nearly three hours." I climbed out of my seat and jumped down to the ramp from the open front cargo door. With my flashlight, I illuminated the wing, propeller, and walked around to the tail. Confirming our cockpit observations, I entered the cabin and noticed our five passengers, a family of four and one college student, had boarded.

"Where did you come from?" Julie asked as I passed her in the aisle.

"I was outside checking for ice. I see our folks made it."

"How's the weather in Marquette?"

"Typical Marquette weather. Snow." I disappeared into the cockpit and informed the captain of my observations. The ground crew completed their duties, and we fired up the engines. We accelerated down the runway and climbed. It was my turn to fly. We broke through the thin overcast and were cruising in the clear, this time toward the northwest. I was hand-flying the leg, as was my habit. We reached our cruising altitude of 10,000 feet and contacted Minneapolis Center, reporting a smooth ride with clear skies above. Negative ice on the climb out of Traverse City. Dan requested a heading toward Marquette until we could receive their navigation radio, thereby cutting a couple of minutes off our filed flight routing, which called for a slight dogleg over Escanaba, Michigan. That was approved.

As we closed within 60 miles of Marquette, Captain Rodts obtained the weather briefing and was told plows were clearing the runway. Marquette measures its annual snowfall in feet rather than inches, and sometimes snow falls for days at a time. It was common to see piles of snow in the woods in early June or for the first snowfall to arrive in September. The previous night at Marquette, snow and ice covered the ramps, and snow was piled in the parking lot as high as the terminal building.

We also heard from our Flight 4227, which also was en route to Marquette, our northern maintenance base, with passengers from Chicago via Green Bay. The first flight into the area normally caused a slight delay to the other because the controllers only allowed a single aircraft on the approach. We were ahead of Flight 4227.

We contacted the controlling facility for Marquette, the military radar facility at the K.I. Sawyer Air Force Base located to the south of town. Our destination lay west of the city, and we received our descent clearance to 3600 feet. I told Dan we would stay above the cloud deck as long as possible to avoid the ice we knew was there. He agreed and also activated all of the anti-ice systems on the aircraft, including advancing the propeller speed to 86 percent. I shallowed out the rate of descent, and we entered the clouds to the south of the airport at about 4500 feet, with the deicing boots at level 3 ice protection selected before entering the tops of the clouds.

Level 1 anti-ice systems operate at all times in the ATR and protect the instrument probes and windshields with electric heat. Level 2 anti-ice systems operate in atmospheric conditions conducive to the formation of structural icing and protect the engine inlets with pneumatic boots, the propellers, and side cockpit windows with electric heating. Level 3 anti-ice systems function when structural icing is encountered and operate the black deicing boots on the leading edge of the wings and tail.

We immediately began picking up moderate rime and mixed icing, and Dan advanced the propeller speed to 92 percent to facilitate ice removal from the blades. We both constantly monitored the ice accretion, and it appeared the boots were doing a good job of removing the buildup, rapid as it was.

"Why don't you tell them I'll take a close turn on," I suggested, meaning I wanted to be radar vectored to the final approach course as close to the outer marker as possible to minimize our time in the ice.

"Good idea," Dan responded and then called Sawyer: "Simmons 4165 would like a turn on just outside of PEEDS. We've got moderate rime and mixed ice in the clouds, negative turbulence."

"Roger, Simmons 4165," came the response. "How about two miles outside of the marker?"

Dan looked at me, and I responded, "That'll be fine."

"That will work for Simmons 4165; please advise 4227 of moderate icing and negative turbulence in clouds," the captain reported to the controller.

"I'm going to fly the final at about 130 or 140 tonight due to the ice," I informed Dan, adding about 15 knots to our calculated reference speed. "I want to configure outside the marker." At this point, we had been building ice for about five minutes.

"Simmons 4165, turn right to heading of 0-5-0, maintain 3600 until established, cleared for the ILS runway 8 approach to Marquette," came the instruction from the controller. "Contact Sawyer approach on this frequency with cancellation and down time or if unable, through Green Bay radio. It looks like you should join about a mile outside of PEEDS."

Things happened rapidly after that. Four years' less experience, and I would have preferred a longer final approach to get everything completed without distraction, but with 3000 hours of experience in the aircraft, this was routine. I asked Dan to arm the approach mode on the flight director, so I would have a backup to my hand-flying of the approach. The two magenta lines on my video display gave me targeted guidance based on computer sensing of the minute electrical impulses generated by the ground-based instrument landing system (ILS) equipment. It was like flying down a chute in a video game.

As the course needle began moving in from the full-left deflection, I preempted the anticipated command of the flight director and asked Dan to turn the heading bug to the inbound course of 79 degrees, simultaneously reducing power further; as the aircraft decelerated to 185, I called for flaps to 15. As we slowed to 180 knots, the computer allowed the flaps to extend, and I retrimmed to remove the changed control pressures. At 170 knots, I called for gear down, and Dan announced "loc star, slope star" meaning the flight director had captured both the localizer course and glideslope.

Promptly at 150 knots, I requested "flaps 30, before landing checklist." Dan pushed the propeller levers slowly to 100 percent, performed the couple of remaining items on the checklist and asked whether I wanted to wait on the landing lights. "Go ahead and turn them on, it won't bother me." The lights came on, and we were in snow showers. The entire cloud was lit up by our landing lights, and we passed over the outer marker on our final approach. At 5 miles from the runway, we were steady at 135 knots and in the chute.

"There's PEEDS," I announced. "Looking for 1606, 200 feet on the slope," referring to the lowest allowed descent altitude without outside references and the associated height above the terrain at that point on the glideslope.

"Right," Dan responded. "Altitude checks. On the loc, on the slope. Before landing checklist is complete."

"Thanks." We had flown the same approach the night before, so it was familiar to both of us. The only difference was that last night Dan was the flying pilot, and the autopilot was flying the airplane. Tonight I continued to hand-fly. We had no turbulence, not even the normal "Marquette chop" caused by the winds moving across the hilly terrain of the area. We knew there were few ground lights on the approach course. We were less than two minutes from touchdown.

We were still picking up ice, but it had begun to change. The accretion on the leading edge of the wing began to split and assumed a "double-horn" formation. The block of ice on the probe outside my window began to build like a rectangular donut. We were now about 1000 feet above the ground. I anticipated seeing the approach lights in less than one minute. I looked back at the wing again. Dan seemingly read my thoughts and made a report on his observation of the ice and effectiveness of the boots. "The boots still look to be working fine. Change in the ice but no evidence of freezing rain. The side window is clear. Less than 1000 feet above the ground. Stay inside and I'll watch for the lights."

"Okay." Although all pilots cannot help but look for visual cues, my main responsibility was to fly the aircraft. I would peek when the time came. The aircraft was 600 feet above the ground, the reported height for the cloud bases. It would be good to be on the ground; we expected to be in the clear soon. I hate the ATR in ice—not light ice, but the moderate icing we see so often. Flight 4165 was about one minute from touchdown.

"You need some rudder trim?" Dan asked. I looked at my "ball" on the turn and bank instrument, which would have shown if the aircraft was skidding slightly sideways through the air. The ball was in the center, maybe one-quarter ball out to the right, but not that he needed to make that call. I was confused by his statement, but I put a couple of quick clicks of trim in to avoid a discussion. The airplane had been stable; we had the power needed to maintain speed; and the glideslope had not varied. It was like riding a train down a track.

Within several seconds, events changed dramatically. The normal sound of the wind passing over the aircraft stopped—not lessened, not changed—but stopped. I felt a slight settling and noticed a slight movement of the aircraft below the glideslope. I added about 10 percent power and then another 10 percent as the excursion began to increase. Without any warning, an oscillation built in the control column, and within a second it had increased to a powerful, rhythmic vibration. I held the wheel tightly and saw we were below 600 feet above the ground and still in the clouds.

"Let's get the hell out of here!" Dan commanded, simultaneously moving both power levers to the stops. I knew to avoid any quick movement of the controls. I realized we had some sort of problem with the tail, but at that moment, it was anomalous and disturbing. Carefully, I pitched up slightly to 5 degrees nose-up as the aircraft began to climb with the massive addition of excess power. A warning, indicating an overspeed situation, began sounding, and I verified my airspeed as 135 knots, well within limits.

"I have 135 knots," I said. "I am going to avoid any big pitch movements." I was acutely aware of everything. The aircraft did not feel like it was accelerating or even in motion. We seemed to hang motionless in the illuminated cloud. A howling wind noise that seemed to pulsate joined the cacophony of electronic warnings, mechanical pulsing of the controls, and increased noise of the engines.

My frightened mind focused on the fact I was not wearing regulation socks. They were going to know I had argyle socks on.

"I have 138 knots on my side. Disregard the warning," the captain commanded. "How is it handling?"

"Okay for now." I reached over and attempted to push the throttles farther forward. My flight instruments told me we were climbing rapidly. I raised the gear to reduce drag. I could control the pulsing control column with one hand, but I held on with both. The oscillations were not moving the controls much because of my grip, but they were very perceptible. I calculated they would move the column full travel if I were to let go, but that was the last thing that was going to happen. Within a minute and a half, we popped out of the top of the overcast. My thoughts were screaming "ICE!"

It was comforting to be out of the accretion. We could not add to what we had. The problem was how to shed what we did have, and what effect that would have on the aircraft. We climbed and accumulated valuable altitude. We verified our indicated airspeed several more times. Each time the speeds on all three indicators were within 2 or 3 knots and always less than 140 knots, well under the speed that should activate the overspeed alarm.

At 7500 feet, we leveled off to see what the airplane would do when we raised the flaps. I cautiously reduced the power from 100 percent as I began to slow the rate of climb so as to not exceed operating limits on the airplane. We were regaining composure, and Dan informed Sawyer approach, which undoubtedly had seen the missed approach on its radar screen and must have wondered what was occurring.

As the aircraft accelerated to 150 knots, Dan raised the flaps from 30 to 15 degrees after we agreed that if I felt anything wrong, he would immediately place them back. The buffeting lessened slightly as the flaps were retracted. I continued the acceleration toward 185 knots, and Dan moved the flap handle to the zero position. I felt the slight pitch changes transmitted through the control column, but the control oscillations moderated more. I continued to gain speed, up to the maximum operating speed of 250 knots. The sensation through the wheel became a high-frequency buzz, and the aircraft was finally stabilized.

I think I took my first real breath since the whole incident began less than five minutes earlier. Convinced that the aircraft control was no longer in jeopardy, the captain contacted the Marquette station and told them that we would proceed to Green Bay. Neither of us wanted to try another approach to Marquette tonight. Dan obtained Green Bay weather, and it was similar to Traverse City, with broken clouds and a temperature of 33 degrees. It made sense to all, and I cautiously turned south.

Simmons Flight 4227 was apprised of our problems and elected to land in Marquette. The crew was based in Marquette; for them it was home. They stayed out of the clouds as long as possible and then made a rapid, last-minute descent to capture the glideslope. This unconventional and discouraged method was not new to Simmons Airlines operations, but textbooks and pilot training warned against this approach. The Federal Aviation Administration (FAA) also frowned on it. Captain Duane Hastrich and First Officer Chris Nielson of 4227 made it home. Captain Hastrich, formerly of Simmons flight management, also realized another ATR icing incident needed to be added to the list.

The landing in Green Bay was on runway 6 toward the northeast. I carefully monitored the changing control forces as the flaps lowered into place, ready for the onset of another buffet. Fortunately, 45 minutes of flight in clear air had sublimated most of the ice formations on the tail, and we made a normal landing in Green Bay. When we taxied to a stop at the gate and shut down the engines, I slid my seat back, leaned forward, placed my elbows on my knees, took an audible deep breath, and turned my head to look at Dan. We had come close, and we knew it. We also knew someone had to do something about the aircraft now.

"I want to get up in their deice truck and take a look at the tail," I said. "Just to see if there is some ice left up there."

"Yeah, I think we should do that. We'll have them get it around as soon as they can. I better call maintenance control and see what they want to do about the airplane. I don't think we exceeded any parameters, but I think they should still check."

Captain Rodts and I exited the cockpit. I turned to close the smoke doors and took a long look at what I really felt was nearly the instrument of my demise. I had never been so scared in an airplane before.

We had a few words with Sue and Julie, giving them the gist of the reason we were in Wisconsin and not Michigan. We wanted them to know some exact details and also find out if they could add anything to the story. They went outside and into the terminal, and Dan followed, placing his blue cap with the silver braid on his head. I was the last crewmember off the aircraft, and I paused to look up at the horizontal tail. Some white protuberances were visible on the upper area of the stabilizer, and I came down the stairs to point them out to Dan, who was 15 feet away looking at the entire aircraft. We went inside to talk with the operations agents and make some calls. It was near midnight. A ramp attendant came into the room and asked, "This is weather-related, right?"

Dan was already on the phone, and I had him repeat the question. "Why?" I said.

"Because if it is weather-related, we don't have to get them a hotel room or anything."

"Listen," I said. "We almost put those people into the earth tonight. You get them a hotel, transportation, meals, whatever they need. And if this company refuses to pay for it, I will." I looked at Dan as the grumbling ramp attendant left the room, and I chuckled. "I cannot believe this damn place."

It was an hour before I got up in the bucket truck to begin my inspection of the airplane for residual ice. A detailed report was drawn up for the company officials and turned in on our return to Chicago O'Hare. I saw no full chord accretion on any of the surfaces or on the fuselage that might have indicated an encounter with freezing rain. Clear ice was evident on many leading edge surfaces, including areas behind the boot and in the aerodynamic gap of the horizontal stabilizer. I also noted a milky-white, spiky formation perpendicular to the airflow on the upper surfaces of the vortex generators on the vertical stabilizer that was never explained.

After 30 minutes of inspection, I climbed down onto the ramp and discussed my findings with Dan. There were many questions but few answers. The flight attendants were waiting in the terminal, and we were told to get rooms at the motel across the street from the airport. Shortly before 2:00 A.M., as I entered my motel room 60 miles from my home, I wanted to call my wife to tell her what happened. I didn't. Later I discovered that Dan also overcame this desire, and we let our spouses sleep. About 2:30, the phone rang, and I was informed we would be scheduled as a flight out of Green Bay direct to Chicago O'Hare at 7:05 in the morning.

On our return to Chicago O'Hare, we were asked to report to the chief pilot's office to receive our instructions for the reports the company required from us. I remember writing those reports with Dan in an office near the common area of the crew lounge. Larry Blacksmith, another ATR captain, was in attendance and questioned us about every aspect of the incident. Larry was shocked by the occurrence and made statements about the aircraft eventually killing somebody if nothing was done.

Slick airline commercials show smiling, courteous employees who are dedicated to serving the public. Airline mechanics in these commercials state that any of them can easily ground an aircraft if they do not consider the aircraft safe. What they do not tell you is that the airlines are multibillion-dollar profit-making corporations. Employees are often told that making a profit is necessary for the company to prosper and for them to keep their jobs. Professional pilots understand this and know there is a darker side to the airline business. Report a problem to management and risk being labeled a troublemaker. If an experienced pilot cancels a flight because of weather or mechanical difficulties, he or she must be prepared to answer for that decision and expect to be disciplined or even terminated.

We assume risk every day. All rational people realize the inherent potential dangers of travel and make judgments regarding the acceptability of the level of a given exposure. Anytime we place ourselves in motion, we face the possibility of a sudden, unexpected stop.

A driver determines that it is safe to cross an intersection based on observations and lightning-quick calculations. Pilots decide it is safe to fly an aircraft based on an intensive investigation of existing weather, his or her own capabilities, the capabilities of the airplane, and the services and navigational aids available at departure and destination airports.

Airline passengers conclude it is safe to fly if they are placed on board the aircraft. Passengers accept the risk because the airline and the FAA place their stamp of approval on the operation, and passengers believe this giant corporation and government agency have done everything in their power to ensure safety.

Passengers' trust is sometimes betrayed.

2

Halloween day 1994

The house was cool and quiet at 4:00 A.M. as John Droy prepared for a one-day business trip to Indianapolis. Sonia, his wife of 41 years, lay asleep in their bed, and as he silently watched her from the bedroom doorway, John smiled. The Droys had five sons and eight grandchildren. They had conquered many obstacles and shared many experiences, yet their life so far had only been the prelude to their upcoming golden years. John quietly moved across the carpet to kiss Sonia goodbye. For a moment he thought of waking her, but it was only a day trip—not really different from a day at the office. Into her dreams he whispered goodbye, and as John turned to leave the room he spotted her button that proclaimed, "It's hard to be humble when you're Swedish" on the dresser across the bedroom.

To remind Sonia of their latest discussions about her Bible study class, with which he had been helping her, John picked up the button and placed it on the top of her Bible on the nightstand. It was a simple, almost absentminded act that took on great significance and offered certain solace to his wife later that day, but now it was just a reminder of his commitment to help her. John saw it was time to go and meet his younger traveling companion, Rob McMillin. For the last time, John Droy put on his coat and left home.

Across town, Rob McMillin, a young father of two, was ready to go. Rob and John had assured their boss at the Ingersoll Milling Machine Company they could handle this call on Cummins Engine by themselves. John was vice president of sales and marketing, and Rob was a senior sales representative; the pair was more than qualified. As a result, in another home nearby a third executive from Ingersoll Milling slumbered as fate passed him by this day.

Barbara Tribble looked forward with pride to October 31, 1994, because it was the day her eldest son, Garvin D. Senn III, would be sworn in as an attorney to the Indiana bar. While growing up, Garvin was referred to as "Trey" because he was the third Garvin, and all three could be around the house at any time. As an adult, he still answered to Trey and joked that being known as Garvin was one of the consequences of growing up. The legal profession was not a traditional one in this working middle-class family; the closest thing to an attorney in the family was Barbara's grandfather Buck, who was known as a "reader of the law" in Kansas around the turn of the century.

For weeks before her trip to the swearing-in ceremony, Barbara talked of little else with her coworkers at Mobil Oil in Dallas, where her job was to ensure that the global relocation and movement of other employees went smoothly. Barbara had grown up as an Air Force brat, and although the family moved to various bases and Barbara lived for a short time in Puerto Rico, as an adult she traveled infrequently, and never before on a small turboprop airliner.

Brimming with no less pride than Barbara Tribble, Eva Calderon had to wait to congratulate her son, Jose, who was being sworn in at the same ceremony in Indiana. Eva had not been feeling well, and Jose had insisted she remain in Chicago while he and his father traveled to Indianapolis for the event. He assured her that they would be home in time for a celebratory dinner.

At the American and American Eagle terminals at Chicago O'Hare International Airport on Monday, October 31, 1994, passengers saw numerous signs and banners proclaiming November 1, 1994, as the 10-year anniversary of American Eagle Airlines. Halloween was the eve of the grand celebration planned for the 10th jubilee. All American Eagle hubs, including Chicago and Dallas, had festivities scheduled for the next day, and outstations all over the system were scheduled to have small parties and cake for staff members. There was an air of excitement and anticipation among the members of the American Eagle family.

Operations were normal throughout the American Eagle system in the early morning, with few delayed or canceled flights. A strengthening low-pressure system and storm front was headed for the Chicago area, and, as the day progressed, the weather worsened, and strong gusting winds limited operations at Chicago O'Hare, the world's busiest airport.

In a new suburban home near Round Lake, Illinois, Orlando Aguiar was awake early and preparing for a day trip in his job as an airline captain for American Eagle. Aguiar voluntarily added this day's trip to his scheduled flying time for the month of October to earn extra money. It would normally have been his day off, because he had completed a three-day trip at 7:30 P.M. the night before, but the four additional flight legs would add a couple hundred dollars to his pay for the month.

Before heading for the airport, Aguiar completed a required medical examination. As a condition of his privilege to act as pilot in command of a scheduled airliner, he was required to carry a first-class medical certificate, renewable every six months. From 9:00 to 10:00 A.M., he was poked, prodded, and examined by a physician certified by the FAA and was subsequently issued his certificate without restrictions. The doctor remembered Captain Aguiar's outgoing and friendly personality and later described him as upbeat, self-confident, and looking well-rested.

Orlando Aguiar understood that one critical physical ailment could end his flying career, so he took care of himself and was in excellent physical condition. He was an avid runner who enjoyed working out on his layovers. He ate a healthy diet, followed a vitamin regimen, and drank alcohol sparingly.

During the previous summer, Aguiar shaved his head completely after being chastised by flight managers for his long hair. When I last flew with him in September 1993, he told me how annoyed he was with the chief pilot, who appeared to have nothing better to do than to hassle crewmembers for their dress habits and hair length. He told me that he wanted to shave his head to show them how ridiculous their concerns were. He did. I will never forget the day I first saw his shorn head during the summer of 1994. I noticed him across the crew lounge as he removed his hat, and all I could do was chuckle and say to myself, "That's Orlando." It had started to grow back by the fall. He had a great sense of humor.

As he drove the short distance from the doctor's office to the airport and parked in the American Airlines employee lot, Aguiar was thinking about the upcoming christening of his daughter, Marisa, and how he was going to be able to work out the trip to Florida for the event.

Jeff Gagliano finished his most recent trip as a first officer on an ATR aircraft for American Eagle on the afternoon of Friday, October 28,

1994. He had the weekend off and was busy at his family's business, a horse ranch, preparing for Halloween before his scheduled return to Chicago O'Hare and next trip on the morning of Monday, October 31. Jeff was based at the Chicago hub for Simmons Airlines, which operates as American Eagle, and had been with the company since August 14, 1989. I remember the specific day because I started with Eagle on the same date; we were new flight crewmembers in the same class.

The Gaglianos' Kettle Moraine Ranch is near Eagle, Wisconsin, on the southern edge of the Kettle Moraine State Forest. October was always busy, with groups coming from all over the state to enjoy the haunted hay ride, a Gagliano specialty. As many as 30 people at a time climbed into a wagon, sat on bales of hay, and rode into the heavily wooded, hilly countryside. Jeff often wore his gorilla costume as he drove one of the tractors that pulled the wagons into the dark. Ranch hands popped up in the dark in costumes as characters from the movies *Halloween, Friday the 13th,* and *Texas Chainsaw Massacre,* as well as other nightmare characters. The ranch was fun and so popular that up to 1000 patrons were treated to the adventure in one day, and crowds peaked near Halloween. This year had been no different, and Jeff had his hands full all weekend. On Sunday, 10-year-old Gina Gagliano reminded her older brother of his promise to go trick-or-treating with her.

"Gina, I have to get this work done for Dad," Jeff responded. He hated to disappoint her—Jeff would do anything for his younger sister. "I have to fly tomorrow and there is a lot to be done."

By Monday morning, Jeff was still concerned about the amount of work around the ranch and considered not flying his scheduled trip. He felt apprehensive about the ATR and the onset of weather that produced conditions that lead to icing. Jeff had been talking about giving up flying and working full-time on the ranch. Maybe it was time.

"Don't worry about it, Jeff," Al Gagliano said. "It's Halloween and we have everything covered for today. After tonight, everything is downhill." Jeff sat on a stool at the kitchen counter, wearing a comfortable orange flannel shirt and eating chocolate cake for breakfast, while his dad leaned against the cabinets near the stove, sipping another cup of coffee. On days that his son was home in the morning before he went to fly, Al usually left late for his job of selling custom-cut meats to restaurants throughout the Milwaukee area, so they would have time together to talk. Al always asked Jeff about his schedule of flights.

Cindy Gagliano hovered about and teased Jeff about his tousled hair and his misbuttoned shirt. Cindy had insisted on a family meal together the previous night to signal an end to the frantic month of activity at the ranch. For the past several weeks, the four Gaglianos had to grab whatever they could or eat from buckets of chicken at odd hours. The meal broke the accumulated stress.

"I packed you some cookies for your trip," Cindy said. "They're on the counter." Jeff usually brought cookies, brownies, or muffins for the crew on the first day of a trip; it had become a well-known tradition among the crews.

"Jeff, are you going to be here for Thanksgiving this year?" Cindy asked.

The young pilot hesitated and cocked his head before answering, but from his body language his stepmother already knew. "No, I won't be here for Thanksgiving."

"That's all right, we'll have the family dinner later in the week like we did last year." After more than five years of Jeff's flying, Cindy understood the disappointments. "How about Christmas? You will be home for Christmas."

"I don't think I am going to be here for Christmas."

Later, the Gaglianos went in separate directions; Al to work, Cindy to do chores around the house, and Jeff down to his room to get ready to leave for the airport. Jeff Gagliano had to drive more than an hour from his home in Eagle to Chicago O'Hare; this morning, with his 10:39 A.M. sign-in time, he had to leave the house before 9:00. Near that time Cindy heard the door close and rushed to the patio door to say goodbye, only to see Jeff's car headed down the gravel lane toward the main road. As she turned to go back into the kitchen, Cindy noticed her stepson had forgotten his carefully wrapped package of cookies.

About the same time Jeff Gagliano left for Chicago O'Hare to start his trip, one of his good friends was already on the third leg of his flying day. Tony Zarinnia was a first officer for Skyway Airlines, the regional affiliate of Midwest Express, and he sat in the right seat of his 19-seat Beechcraft 1900 airliner at the gate next to American Eagle's gate in Indianapolis, waiting for passengers so the flight could depart for the return to the airline's hub in Milwaukee. Indianapolis was chilly, and a light, misting rain fell from the gray sky. Tony sat in his airplane because the first officer of a 19-seat commuter airliner is also the flight attendant who stands outside as the passengers board, and he wanted to stay dry as long as possible. Tony had already spent too much time standing in the rain in his aviation career.

The Beechcraft 1900 had been unable to climb above a line of weather between Milwaukee and Indianapolis on the flight south, even at 23,000 feet. Ice was there, and during the flight, Tony was reminded of what Jeff had said about how badly the ATR handled ice. Even when Tony had been flying the light twin-engine Barons running night freight or checks, Jeff said how Tony was better off in the ice than Jeff was in his heavy, turboprop airliner. That type of reverse efficacy is almost unheard of for aircraft; usually the larger the airplane, the more capable. Icing was such a concern to Jeff Gagliano that he had told people he would die in an ATR because of ice. As Tony spotted passengers walking out into the mist, he pulled himself out of the seat and stepped down onto the pavement to greet them.

Patricia Henry and her two sons, Jonathon and Patrick, woke on Monday morning in a hotel in Indianapolis. Taking a friend along to watch the boys while Patty worked, the family had traveled on an excursion that combined a getaway for the boys with some business for Patty. As the president of U.S. Intell and a licensed private investigator, Patty had her choice of assignments, and on the previous Wednesday, October 26, a client telephone call started them on this journey. The company usually worked for large commercial clients across the country, but something in the voice and story of the woman on the other end of the telephone touched Patty. Although the company declined to take the woman's case at first, the decision was reversed after more thought. Patty determined that she was the best equipped for the mission required by the client and decided she and the boys would fly from Chicago to Indianapolis on Friday.

With Jonathon, seven, in school, Patty had taken four-year-old Patrick to the U.S. Intell offices on Friday while she confirmed her United Airlines tickets and made other phone calls. While his mother was on the telephone in her office, Patrick played a game with Terri Severin, Patty's sister, as she worked at her desk. When his aunt turned away, Patrick snuck up to her desk and stole a small portion of the snack she had laying there and then ran to his mother's side. After seeing him go, Terri "noticed" some of her food missing and looked all over and under her desk for what had disappeared, to the delight of the small thief. Untiring of the game, Patrick snuck back every time Terri turned away her attention until his mother finished her phone calls.

Patty, Terri, and Patrick left the office and stopped at Terri's house before mother and son left to pack, pick up Jonathon and the friend who would travel with them, and get to the airport for the afternoon flight. The sisters talked and Patrick played with his favorite toys—miniature cars and a parking garage. Terri and Patty were close, possibly because they were "Irish Twins," born on the same date one year apart. The pair had a knack for unspoken communication, and a friend said it was difficult to tell when one left off talking and the other started when they were together. When time came to leave, it took both sisters a few minutes to convince Patrick that he could play with the cars and garage the next time he stopped over.

On Sunday evening, Terri and her husband, Tom, had invited Terri's parents over for dinner. During the meal, Patty called, but before Terri could talk to her sister, the call disconnected. Patricia did not call back, and Terri did not know which hotel they were staying at, so she could not call Patty.

On Halloween, Patty called her mother and said she had stayed an extra day in Indianapolis to do some more work on her case. Before hanging up, she told her mother that she and her sons would be home in time to go trick-or-treating.

Cheryl Dwyer was also looking forward to Halloween, and she had agreed to the business trip to Indiana only with the understanding that she would be home in time for children to come trick-or-treating. The request from a client of Hewlett-Packard, her employer, to demonstrate some medical equipment manufactured by the company had originally been assigned to one of Cheryl's coworkers. When the other employee could not make the trip, Cheryl reluctantly agreed to go. Cheryl and Dennis, her husband, had moved into their new home in Arlington Heights, northwest of Chicago, during the past summer, and this would be their first Halloween in the new neighborhood. Anticipating the festivities, Cher purchased candy and placed it on the table. Delighted because she finished her business appointment early, in the late morning, she rushed back to the airport to try and catch an early flight; the ticket agents were unable to fulfill the request, so she waited to fly on Eagle Flight 4184.

Jeffrey Lietzan and Alison Smith-Field had met at an American Eagle gate at Chicago O'Hare International Airport a couple of days before

Thanksgiving 1993. Jeffrey had recognized the pretty young lady taking a seat in the waiting area as a fellow passenger on the flight that had just arrived from Washington, D.C. Circumstances had provided the nervous bachelor with the perfect opening line, and he approached Alison.

"Hi, weren't you just on the flight from Washington?" Jeffrey asked.

"Yes, I was," responded Alison.

"So, where are you off to now?"

"Indianapolis."

They were both on the same flight to Indianapolis. As they sat and waited to board, the pair struck up an enjoyable conversation that quickly passed the 40-minute wait and the one-hour flight. By the time Jeffrey and Alison parted company in Indiana that day, they had found a number of things in common and agreed to be in touch after their return to the East Coast.

On October 31, 1994, Jeffrey and Alison were in Indianapolis again. They had to be at the airport to catch their flights home by early afternoon, and there was so much left to be done. A stop at the courthouse to secure a marriage license, a dress fitting at Alison's mother's house, a meeting with the minister, and several calls on the final details filled the morning hours just as similar chores had filled the weekend. Best of all, they were together, planning their wedding for the Saturday after Thanksgiving, just one year after they had met.

The young couple had used flight coupons and frequent-flyer miles to save money. With the upcoming wedding and the expenses of starting a life together, they could use the savings. They were taking different flights. Jeffrey was flying directly from Indianapolis to Washington, D.C., and Alison was following via a connection in Chicago after a short flight on an ATR-72—American Eagle Flight 4184.

A knack for music and an understanding of engineering are a rare combination, but Brad Stansberry was a rare person. In his job as a project engineer at Delco Electronics, a division of General Motors, he was able to combine both talents by designing audio systems for General Motors. His latest assignment had taken the 1989 Purdue University graduate to the Opel plant in Wiesbaden, Germany. Living in Europe was exciting, but the last 10 days at home in Anderson, Indiana, were heaven. Brad enjoyed the time so much that he failed to check the date on his ticket and missed his scheduled flight overseas on Sunday.

His vacation back in the United States could not have been timed better. It was homecoming weekend at Purdue. Brad could attend as an alumnus and accompany his siblings, Jennifer, a senior, and Matthew, a junior. Early Saturday morning found Brad and Jennifer at a 1950s-style, all-night restaurant eating pancakes and talking until 4:00 A.M.

Even though Brad had missed his scheduled flight on Sunday afternoon, American Airlines was able to accommodate him on Monday. Bruce Stansberry drove his son to the Indianapolis airport during the afternoon while the pair talked about Brad's other love—sports. The Indiana Pacers had begun another season and the Stansberrys hoped for a winner this year. It was what Jennifer described as "typical male-bonding stuff."

Three other Delco Electronics employees were traveling on the same flight to connect to international destinations out of Chicago. While waiting in the boarding area, Brad overheard a group of four people discussing how they would have to split up for the travel to Chicago O'Hare. Not wanting to see a family divided for the flight, Brad offered his seat on American Eagle in exchange for one held on the United flight to allow Jonathon Henry to fly on American Eagle with his mother, Patricia, and brother, Patrick. His offer was declined—Jonathon would fly on the United flight with the family friend—and Brad Stansberry boarded the red, white, and blue ATR-72 along with his coworkers.

Dana Thompson had a family tradition in aviation: Her mother had been a flight attendant for American Airlines and her maternal grandfather, Byron "Pop" Warner, was one of the first jet pilots for the same company. Even with the lure of flight in her blood, Dana decided to pursue the lure of another industry—television. Her talent, hard work, and pleasant demeanor had proven her to be a rising star in her profession. Nobody should have been surprised. Dana was successful at anything she tackled. Her work ethic had been established by the time she took her first job selling candy in the Chicago suburb of Glenview. An internship in 1988 led to a job as a sales associate for the USA Network the following year. Within five years of joining the company, Dana had been promoted three times, and another door was opened to her at 20th Century-Fox's "f/X Network," where at 27, she became director of affiliate sales and marketing for the central region of the new cable network. It was in that position that she found herself traveling from her office in Chicago to Indianapolis.

"The entire next week is going to be hectic with travel," Dana told her father and paternal grandmother during their first visit to her new apartment in Chicago the Saturday before Halloween. It was the first time Buddy Thompson, Dana's 88-year-old grandmother, had felt well enough to see her granddaughter's new home, which meant a lot to Dana. She had been giving her grandmother ceramic angels as gifts since she was a child, and Grandma Buddy had always displayed the treasures with pride. Dana was close to her family and often referred to her dad as her best friend.

Her father, John Thompson, had been thrust into the role of single parent by his wife's death when Dana was five years old. "I was a single parent before it was fashionable," John said. He was proud of Dana's accomplishments. The previous Labor Day, she had taken her father to London for a week using frequent-flyer miles, hotel coupons, and her own money to thank him for everything he had done for her. When he remarried 10 years after the loss of his wife, the relationship between Dana and her father had flourished with the addition of Kay Thompson to the family.

Dana was flying from Chicago O'Hare to Indianapolis to visit customers on Monday morning and flying back in the afternoon so she could attend the Chicago Bears-Green Bay Packers football game with coworkers.

Bruce Anglemyer kept his priorities straight, and the people he worked with at Fast Food Merchandisers admired this quality. As treasurer of the company that services the Hardee's fast-food chain, the 49-year-old father of two often walked up to late-working coworkers and said, "Go home; this will wait. Take care of what's important." In the high-pressure corporate world, that philosophy enlisted substantial employee support and goodwill.

On Friday, October 28, Bruce traveled from his home in Rocky Mount, North Carolina, to Indianapolis to perform an inventory audit at his company's produce facility. Working the rest of Friday and all day Saturday, he finished early enough to make a short trip to northern Indiana to visit his mother and stepfather for a couple of days before his return home. Bruce arrived in South Bend late Saturday, attended church on Sunday and then went to brunch. His 78-year-old mother recalled how her son had made her laugh by commenting on the small jelly packets: "When I retire, I am going to buy an orchard and grow all these different kinds of apples and put them in these little containers."

That sense of humor had helped Bruce win the heart of Sandy, his wife of 25 years. Together they had raised two sons, Brad and Jon, and they looked forward to the upcoming years on their lake property, on which both dreamed of building their retirement home. Sandy wanted the home to have a view of the sun as it set over the channel that bordered their property and a porch swing on the dock so they could watch the day fade behind the trees.

Monday morning, Bruce said goodbye to his mother and began his four-hour drive back to Indianapolis to catch his flight to Chicago and then home. He could have booked his flight with USAir, which would have meant a shorter route with less backtracking than the American Eagle route, but Bruce was concerned about recent problems, including a crash, at that airline. What had seemed like a good idea when he booked the trip now made no sense to the CPA. "I will never do this again," Bruce told his wife during a phone call over the weekend.

Anthony Moore had learned that society could be enriched by the cultivation of its young people. His mother, Connie, is director of the YMCA Urban Youth Services Program at Michiana in northern Indiana, and Anthony had been working with the youth of Seattle in a similar program for 18 months. Anthony was home in Indiana to participate as a groomsman during the weekend wedding of a friend's brother. The short visit also gave Anthony some time with his fiancée, and the two finalized plans for her move to Seattle the following week.

By sitting up until 2:00 A.M. talking with his brother and then rising early on Monday morning to travel to Indianapolis for his afternoon flight, Anthony had accomplished much during the mini-vacation. As tiring as the hectic period had been, Anthony still had plenty of energy left as he bid his friends and family goodbye. As people had grown to expect, his smile was contagious.

David Shellberg had loved meteorology since he was a small boy. He made weather observations and created elaborate logs that he updated daily. Unlike the itinerate nature of many of the passions of children, David's interest in the study of weather became more focused and intensified after an encounter with a tornado when he was seven. His mother had wanted him to come to the basement with her for shelter, but the brown-haired boy with the inquisitive eyes had wanted to see and learn. That pursuit of knowledge had placed 25-year-old David at the head of the table in the meeting room at the Indianapolis station of the National Weather Service, giving a presentation on an automated weather system known as "Mesonet" during the late morning of October 31, 1994.

Few individuals were better qualified than David to talk about the system that he had been working to develop and install in Oklahoma. The old hands at the National Weather Service were impressed with the system's performance, which placed 110 remote data-gathering units throughout the state and provided real-time information and 24-hour recording of the weather. Because weather can change rapidly and often, meteorologists follow the adage "the more data, the better the guess." The observers were also impressed with David's confidence and pleasant demeanor.

In David Shellberg's perfect world, the Weather Channel was on every channel; everyone had plenty of chocolate-chip cookies; and his three favorite teams, the Detroit Pistons, Detroit Tigers, and Purdue University Boilermakers, won every game. It was the Purdue homecoming that had brought David from Oklahoma to Lafayette, Indiana, 30 miles northwest of Indianapolis. He participated as a member of the Alumni Band at the football game. A near-record 340 alumni returned to play with the band at the game; the oldest was 70. The game with Iowa ended in a tie, but the band had the stadium on its feet and cheering during the halftime show.

David mentioned the homecoming-weekend experiences to some of the weatherpeople attending his presentation. Some later recalled how he had told them that he had stayed in the area longer so he could introduce Mesonet to the Midwest, and they seemed unanimous in their high opinion of the system's value.

David Landes was a Kappa Kappa Psi and another Purdue alumnus who performed with the Alumni Band during the halftime festivities this homecoming weekend, but he did not have to travel far. He lived in Indianapolis with his wife, Sally, and had begun a new career in September as a customer-service representative with Physicians Practice Management after six years of teaching mathematics to junior-high students. On Halloween, he traveled to Rockford, Illinois, for business. With coworker Jerry Thomas, David waited in the American Airlines departure area for their flight to be boarded as David Shellberg approached the ticket counter to check in.

For Kenneth Spencer, flying was just another part of modern life—as safe as driving to the store to buy milk. The statistics favored passenger safety aboard an airliner. With his wife, Roberta, Kenneth had returned from a weekend business trip to Modesto, California, the previous day and changed a ticket on a flight from Midway Airport to the more convenient larger airport because he preferred flying with American.

As an experienced traveler, Kenneth displayed a familiar prejudice that surveys often confirm. Despite discount fares, frequent-flyer gimmicks, and slick commercials, polls have shown that people often prefer a particular airline for reasons that date back to their first airline flight or the line that served their small hometown airport. Familiar paint schemes and airline names provide a certain comfort level to passengers, and many hours are spent researching the subject before executives decide to modify a corporate logo or change a paint scheme. American Airlines understands that recognition and brand loyalty and jealously guards its silver airplanes with the red, white, and blue stripes. The company had endured internal struggle on par with a hostile takeover attempt when someone had suggested changing the shape of the eagle logo on the tail in favor of a more updated, stylized version.

That same type of comfortable familiarity is recognized in the attempt to provide a seamless level of service among American's regional affiliates. Four separate companies operate the American Eagle system, and even though flight attendants wear similar uniforms and the seats are upholstered with the same fabric and pattern as the coach seats of American Airlines, the ticket purchased by Kenneth Spencer for his trip to Indianapolis never took him near an American Airline plane. The ATR-72 he boarded was operated by a crew trained in comparable simulators to the standards of American Airlines and operated under the same regulations. The turboprop aircraft spent much of its time, however, in heavy weather, and its abilities in icy conditions were poor compared with the United Airlines Boeing 737 parked on the other end of the concourse. It made financial sense, though, to AMR, parent company of American Airlines, to use affiliates for some routes.

Colombian toymaker Semmy Grimberg never flew on the same airplane as his 24-year-old son, Adrian, to ensure the survival of the family business should something happen to either of them. They traveled frequently, and they had talked about the travel arrangements for this particular trip before leaving their home in Cali. Semmy had expressed his concerns about flying together, but this flight from Indianapolis to Chicago would be short. Finally, Semmy relented and agreed they would travel together, accompanied by Semmy's wife, Guilda, to the Industrial Fabrics Association International Meeting that was to conclude with a banquet on Tuesday night. The trade show had been a successful excursion for the Grimbergs and they were leaving on Monday on American Eagle.

Six other attendees from the convention boarded the flight with the Grimberg family, and like the Grimbergs, some came from other countries. Wan Suk Ko was on his first visit to the United States from Korea and boarded the flight to travel to Chicago and visit his brother. Pierre Bonneau came to the meeting from Canada and was flying home to his wife and two children. Nancy Baker and Jay Ganong arrived from southern California and were heading back to hospitable weather. Gino DeMarco was making the shortest trip home; he lived in Mount Prospect, Illinois. He came to Indianapolis to meet with officials of the association for only three hours on Monday. Gino called his wife, Lourdes, before boarding the flight and told her he would be home about 4:30 P.M. Being home early was important to the soon-to-be first-time father. He was practicing his parenting skills on his young nieces and nephews and especially wanted to see them in their Halloween costumes.

The last conference attendee to board was Frank Sheridan. Frank's business manufactures covers for spas. The Navy veteran had earned three Purple Hearts during his service in Vietnam, enjoyed scuba and skydiving, and built and flew his own planes as a licensed private pilot. The long-time Californian wanted to ensure that he made his connecting flight to the West Coast at Chicago O'Hare and requested to be placed on the standby list for Flight 4184 even though he was scheduled to be on the next flight.

Passenger Maurice Stein provided long-remembered, special experiences to disadvantaged youngsters. With his wife of 30 years, Amy, 58-year-old Maurice operated Camp Echo Lake near Lake George in Warrensburg, New York. Just as the Gagliano family's Kettle Moraine Ranch invited children with disabilities or children from the inner city to Eagle, Wisconsin, Morry and Amy Stein invited about 55 kids each year for a free week at Camp Echo Lake. As the chairman of the board of the American Camping Association, Maurice had been in Indiana canvassing for support and donations in the final drive to reach the foundation's $1-million goal for its endowment fund.

Sandi Modaff had wanted to be a flight attendant like her older sisters, Betsy and Kim, and when she was hired by Simmons Airlines in January 1987, she became the third of the four girls in the family to become a flight attendant. The remaining sister, Missy, joined the airline a short time later. Family was important to Sandi. She was one of those special aunts everyone wished they had. She sent cards on special occasions to her sister Kim on behalf of Kim's daughters, Brooke and Courtney, signing their names.

Friends also were important to Sandi, and in training she met many people who became part of her life. When she was hired, flight-attendant training was held in Marquette, Michigan, on Lake Superior. The boats of the Great Lakes captured Sandi's fancy, and to her they represented new experiences and places. American Eagle opened opportunities to the 21-year-old woman, and everywhere she went, she collected memories and left smiles.

Early on the morning of October 31, Sandi talked with her mother, Carole, while she was in the kitchen packing lunch for her trip. In typical flight-crew fashion, her meal was unbalanced—lasagna and popcorn.

"Sandi, the weather is really bad," her mother said. "It's very cold, windy and rainy."

"If the weather is too bad, they'll cancel the flight, Mom."

"That would be all right with me. Then you'd be coming back home soon."

"It usually doesn't work out that way. Sometimes you get delayed or stuck someplace when they cancel the flights. I've flown in weather much worse than this—blizzards and ice storms. Mom, don't worry."

With that, Sandi was off to the airport. She looked forward to the week, because she was planning a reunion dinner on Thursday with her long-time beau, an ATR captain, and had packed a special dress for the occasion. For the first few days of the month, the flight schedule was in transition, and she was flying with reserve captains and her regular first officer for November, Jeff Gagliano. By mid-week, however, her favorite, "Captain Paul," would be back. On this day, though, she smiled when she learned she was flying with Orlando. He always made her laugh.

In the operations area, just after sign-in time, the crew coordinator approached Sandi and Captain Aguiar and introduced a fresh face—Amanda Holberg. Today was one of the first line flights after training for the new flight attendant from Texas, and Sandi welcomed her warmly; there was always new friend potential in everyone she met. The three prepared to leave for the airplane about the same

time. Designated N401AM, the plane was one of the newer ATR-72s
in the Simmons-American Eagle fleet.

American Eagle flight attendant Jeff Shelton always seemed to be
smiling and joking, and he could hold a conversation with just about
anyone. Everybody knew Jeff. While making his way through the
crowded Chicago O'Hare crew lounge around 11:00 A.M., he spotted
his friends Orlando Aguiar and Sandi Modaff in the operations area.
He later recalled his last conversation with them:

"Hey, guys," Jeff said with a big grin. "How's it going?"

"Great, man," Captain Aguiar smiled back. "I get to fly with this
great lady today. How's that for being lucky?"

"You always catch the breaks. Are you two flying together the
whole month or is this just some kind of transition thing?"

"I just picked up part of the day, man. Going for another hundred
hours of pay this month. Don't forget I got that new baby to start sav-
ing for."

"Where are you guys headed?" asked Jeff.

"Indy, for the big sit," replied Orlando.

"Yeah, well you better get used to it," said Jeff. "Weather is bad
today, and you may get a bigger sit than you realize. My crew was
saying it may not be too much longer before they just shut things
down for the day and leave people wherever they are for tomorrow.
If they do that and I come to Indy a little later, we could get together."

"Yeah, right." Orlando grabbed his paperwork off the counter
and headed for the door. "Later, man."

"Catch you later," Jeff called. "In Indy."

Orlando Aguiar and Sandi Modaff were still on the ground in Indianapolis, awaiting their release back to Chicago O'Hare when Jeff
Shelton's flight arrived in Indianapolis. The pair were aboard American
Eagle Flight 4184 waiting on the ramp. Orlando and his first officer
watched the landing of Eagle Flight 343, but the crews never met up.

John Knapp's friends dubbed him "Crankshaft" after the comic-strip
character with the rough exterior but soft heart. John's resemblance to
his fictitious counterpart went beyond physical appearance—
both are school-bus drivers. The cartoon Crankshaft is rough-shaven,
often ill-tempered, but he has a heart of gold that peeks out on occasion. The flesh-and-blood Crankshaft is rough-shaven and has the appearance of someone who could have an ill temper, but his quick, easy

smile and sincere eyes showed his deep feeling and true caring nature. John Knapp was loved by the children on his school bus, who ran smiling to board the yellow transport and often exited only after providing a big hug for the grandfatherly figure.

John also was a long-time member of the Lincoln Township Volunteer Fire Department, for which he organized specialized training. During the months before this Halloween day, John noticed the streams of air traffic that passed over Roselawn en route to Chicago's two major airports and as a result had made 15 to 20 calls to the FAA during the summer and fall of 1994 requesting airplane-crash fire-and-rescue training for the small department. Finally, John talked with the chief of the fire department at Chicago O'Hare International Airport, and it was made clear that policy had precluded the training of departments as far away from an airport as Roselawn. However, Chief Meade had indicated he would look into providing some type of information. As of October 31, 1994, that information was yet to be received by the rural fire department at Roselawn.

My Monday started early in Lansing, Michigan. Our crew was beginning our third day of a four-day trip with the early morning flight to Chicago O'Hare. We were scheduled out at 6:00 A.M., but the actual departure time was two minutes late. The delay did not affect the flight. I was flying with a senior captain and two new flight attendants, and this was our last trip of the month. Unlike most of my coworkers, I liked to fly weekends so I could have a couple of free days during the business week to spend at the offices of my business, and this schedule met my needs. Jeff Gagliano, my classmate, mentioned his appreciation of my choice on several occasions during our five years at the airline. My seniority was a couple of numbers higher than his, and my choice to fly weekends often allowed him a schedule with more Saturdays and Sundays off so he could help at home. I joked he could leave the money in my crew lounge mailbox.

Our flight took us into Chicago a little after 7:00 A.M., followed by a turn to Toledo, Ohio, and a noon departure for our overnight in Cedar Rapids, Iowa. I began to notice the weather on the way into Chicago O'Hare on that first flight of the day. We picked up some ice and encountered some turbulence, but it was on the flight to and from Toledo in the later morning that I began to feel uncomfortable. Earlier in the month, I had attended recurrent ground school. We had discussed the status of icing problems with the ATR aircraft. The ground instructor, formerly of American Airlines, talked about the poor performance history of the ATR, and then he mentioned an icing incident during an approach by an ATR to Marquette, Michigan, 11 months earlier.

"In September I had a captain in class who told me about the most recent ice problem in the ATR," the instructor said. "I wanted to pass along the story of what happened to one of Simmons' flights last November."

"I was flying the airplane when it happened," I said, volunteering to tell the class the details of the event.

"You were with Dan Rodts?" asked the instructor.

"Yeah."

We recognized a specific type of ice during the November 1993 incident, and I saw that same type of ice on our turn to Toledo early on Halloween day. The weather system was intensifying by the time we returned to Chicago O'Hare shortly after 11:00 A.M. I was thinking that this was a dangerous situation as we bounced through the rain shafts accumulating double-horn icing on the aircraft. My eyes were focused on the icing probe, a round piece of aluminum outside of my window, and as a block of ice developed on it, I was frightened by the similarity to the incident the previous November. I overcame my nervousness by reminding myself that lightning rarely strikes twice, and I already had my ATR icing incident.

When we parked on the Chicago O'Hare ramp at 11:10 A.M. on October 31, 1994, I told Captain Steve Henry we should let our company dispatch know about the weather. He agreed and left the cockpit ahead of me. Walking to the crew lounge, I saw Jeff Gagliano across the concrete, doing a preflight inspection on his aircraft, and waved. Farther down the walkway, I passed other crewmembers and nodded to Captain Orlando Aguiar, flight attendant Sandi Modaff, and new flight attendant Amanda Holberg.

American Eagle Flight 4101 from Chicago arrived in Indianapolis at 12:50 P.M. Central Standard Time (CST) October 31, 1994. The new ATR-72 glistened on the ramp, still carrying moisture from the clouds on its shiny red, white, and blue paint. Some passengers bound for Chicago were already in the gate area, and many of them walked to the windows overlooking the ramp to watch the disembarking inbound passengers and inspect their ride. The crew, Captain Orlando Aguiar, First Officer Jeff Gagliano, and flight attendants Sandi Modaff and Amanda Holberg, had a scheduled 80-minute break on the ground before returning to Chicago as American Eagle Flight 4184.

3

American Eagle Flight 4184

At 2:03 P.M. CST, October 31, 1994, American Eagle Flight 4184, scheduled nonstop service between Indianapolis Airport and Chicago O'Hare International Airport, was in the final stage of boarding passengers. The aircraft was refueled and serviced for the flight, and the last of the luggage checked at the gate was loaded into the aircraft by the ground crew.

Captain Orlando Aguiar was a seven-year veteran of Simmons Airlines, which is owned by AMR Corporation and operated under the name American Eagle. Captain Aguiar had 7900 flight hours, 1550 of which were in the same type of ATR aircraft he was commanding for this flight. He was in the cockpit, carefully reviewing the weather and other particulars for the trip. To his right sat First Officer Jeffrey Gagliano, a five-year airline professional with nearly 5200 flight hours and more than 3700 hours in the ATR-42 and ATR-72 aircraft.

Flight attendants Sandi Modaff and Amanda Holberg welcomed people aboard and helped the 64 passengers find their seats and suitable stowage areas for the extra carry-on items that wind up in the cabin. Storage space is limited in the cabin of the ATR-72 when it is full; on this day, American Eagle Flight 4184 carried the maximum number of passengers. Sandi was a six-year veteran of the airline, and her confident demeanor allowed her to take charge of the situation. As the senior flight attendant on this trip, Sandi could choose her cabin position and duties. Each cabin position has its pros and cons, and she decided on flight attendant position 2 and stationed herself in the front of the cabin. Amanda, on one of her first trips as a flight attendant, was eager to please, excited, and proud of her new job.

The weather was still pleasant in Indianapolis; the air temperature was 65 degrees with a 10-mph wind from the southeast. In Chicago, the temperature was a cold 42 degrees, and rain showers and strong, gusty winds blew from the northeast off Lake Michigan. Amanda wanted to transfer her base to Dallas just as soon as possible to be closer to home and to avoid the horrors she had heard about Chicago winters.

"Clearance, Eagle Flight 184 to Chicago O'Hare with Sierra" was the first radio call for this flight made by First Officer Gagliano, recorded by Indianapolis Clearance Delivery tapes. American Eagle Flight 4184 was known to air traffic control as "Eagle Flight 184." "Eagle Flight" identified the aircraft as an airliner operated by American Eagle, and shortening the flight number to the three-digit "184" made transmissions easier for both the crew and controllers. The reply came back almost immediately from the air traffic control specialist, reciting the clearance items in their normal precise order.

"Eagle Flight 184, Indy Clearance, cleared to O'Hare Airport via vectors Boiler, Victor 7 the Bebee intersection, direct O'Hare. Maintain two thousand five hundred, expect one four thousand, one zero minutes after departure. Departure frequency will be one one niner point zero five. Squawk six six five two. You have expect departure clearance time of two zero five two; time now is two zero zero four."

In short, American Eagle Flight 4184 was approved by air traffic control to make the flight from Indianapolis to Chicago O'Hare. To maintain proper traffic separation, the departure controllers provided initial radar vectors to the flight designed to guide them safely out of the Indianapolis area to a point along the airway routing that would take them to Chicago O'Hare. Airways are the highways of the sky, formally referred to as Victor airways. Intersections occur when two or more airways intersect, just like roads on the ground. Each intersection in the United States is identified by a unique five-letter identifier. Sometimes these intersections are named for a town or a landmark or even a sports team in the area of its geographic location. Pilots navigate on these airways by using radio receivers on board the aircraft that interpret signals and display information from transmitters on the ground called VORs, an abbreviation for very high frequency omnidirectional range station.

For this flight, after following the radar vectors from Indianapolis Departure Control to join the airway, the crew would pilot the aircraft northwest on Victor Airway 7 over the Boiler VOR, located near Lafayette, Indiana, home of the Purdue Boilermakers, to the Bebee Intersection and then expect to proceed directly to Chicago O'Hare, once again following radar vectors from Chicago Approach Control. The flight was instructed to climb to an initial altitude of 2500 feet. All altitudes are measured above sea level for continuity. This altitude would keep the aircraft at a safe height and allow the controllers time to identify the flight on the radar screens before permitting it to climb higher and mix with other air traffic.

In case radio communication could not be maintained with the flight for whatever reason, the pilots were told to expect one four thousand 10 minutes after departure. This instruction allowed the crew to climb to 14,000 feet 10 minutes after takeoff and proceed along their cleared routing at a safe altitude. Flight 184 was to contact the departure controller on frequency one one niner point zero five, or 119.05 megahertz, and the pilots were to set their transponder to "squawk" code six six five two. The transponder on an aircraft transmits a code number, in this case 6652, to all ground air traffic control radar facilities. Ground radar picks up the code on every sweep of the radar beam and displays the information as a blip on the air traffic controller's screen. Along with this identifying number, airliners and many other aircraft also automatically transmit the airplane's altitude to radar stations.

The last sentence confirmed Captain Aguiar's concerns regarding delays on the flight; Eagle Flight 184 was not allowed to take off until 2:52 P.M. CST, 49 minutes from the time First Officer Gagliano made the initial radio call and 37 minutes later than scheduled. The choice of the crew to take the delay on the ground at the runway, where they would be ready to go if air traffic control should have an unexpected opening in the flow of aircraft toward Chicago O'Hare, was a smart operational call not only because the crew knew that about half of the time the delay is either shortened or eliminated altogether but also because it starts the pay clock for the crewmembers. Until the flight moves away from the gate, known as "blocking out," the pilots and flight attendants are not paid for their time.

First Officer Gagliano copied the clearance onto the takeoff performance card using the pilot shorthand learned early in his training and perfected during years of flying and obtaining thousands of similar clearances. He read back the instructions over the radio, a requirement of the FAA to make sure that all pilots and controllers are working from the same page of the playbook. In aviation, there is no room for guesswork. Jeff needed no maps to find his way to Chicago O'Hare because he had flown the routing given in this clearance hundreds of times and knew it by heart, including the radio frequencies to expect along the way; however, like any good pilot, he had his charts out and ready for any last-minute changes or diversions. Familiar air traffic controller voices were likely to greet each of the radio calls from the flight to every new radar sector. Although most pilots and controllers never meet each other face-to-face, they come to recognize the voices on the other end of the radio.

In the terminal, Charlew Beyers said goodbye to her mother, Lewise Morris, as the older woman left to get on board the American Eagle flight. Watching the ATR-72 through the window, Charlew noticed an attractive blonde woman with a young child at her side rush up to the ticket agent.

"Is there still time?" asked the breathless young woman.

"Sure is. They let us know you were on your way," responded the agent as she took the tickets and directed them toward the door leading to the stairs down to the ground level. Patricia Henry and her four-year-old son Patrick hurried to the aircraft. Charlew Beyers watched from the terminal window as little Patrick ran across the ramp, followed closely by his mother. At the aircraft stairs, the mother scooped up her son and carried him on board American Eagle Flight 4184. They were the last people to board.

Two hundred feet away from the ATR, a United Airlines Boeing 737 jet, also bound for Chicago, waited at the gate for approval to close its doors. United's flight to Chicago O'Hare was also full; among its passengers were Patricia Henry's seven-year-old son, Jonathon, accompanied by Patricia Henry's traveling companion and friend. They planned on meeting at Chicago O'Hare.

At 2:14 P.M. CST, the ground crew was completing the loading of the final bags into the aircraft cargo compartments, and the lead agent walked across the ramp to the rear entry door of Flight 4184. The lead agent saw Amanda Holberg standing in the doorway, waved and said goodbye, and helped close the aircraft door.

The lead man of the ground crew saw the cabin door close and knew the flight was waiting on his people to finish their jobs. The wind was starting to pick up and nobody wanted to be outside; the crew worked quickly and carefully. As the cargo door behind the cockpit closed, the lead man glanced toward the front of the aircraft and signaled for another of his crew to approve the starting of the engines. Flight 4184 would be out of the gates a few minutes past its scheduled departure time but would still register as an on-time departure for FAA record-keeping.

In the cockpit, the pilots recognized the thud from the rear of the aircraft as the cabin door closed, and both Orlando and Jeff glanced at the overhead panel to watch the cabin door warning light wink out as the system sensed that all of the locking mechanisms had operated properly. Holding up two fingers, Captain Aguiar looked at the ground crewman standing at the nose of the airplane, who pointed to the right-side engine and rotated his other hand above his head. The signal to start engine number two was confirmed, and the crew completed the checklist and fired up the starboard engine. As a result of the hard right turn needed to leave the gate area at Indianapolis, the crew elected to also start engine number one on the left wing to help with the turn. Normally, with a delay of more than 30 minutes and moderate outside temperatures that do not require extra air conditioning, the crew could have taxied to the runway on one engine to conserve fuel and keep the cabin noise lower.

With both engines running and stabilized, the captain would have called for the "After Starting Engines" checklist to make sure the crew did not forget any items before moving the airplane. Everything is double-checked on an airliner with the use of checklists. Although a crew has completed the starting operation thousands of times, it always executes the entire document just to be sure and to comply with FAA regulations.

After the checklist was complete, First Officer Gagliano called the appropriate air traffic controller. His radio call was recorded on tape:

"Indy Ground, Eagle Flight 184 is ready for taxi. We've got Sierra."

At 2:17 P.M. CST, the pilots received further instructions from Clearance: "Eagle Flight 184, Indy Ground, taxi runway two three right. Pull into the runup pad. TWA is also awaiting a release time."

In the cabin, the flight attendants completed their pretakeoff announcements and final checks. Even though she was told about the delay, Sandi was experienced enough to know how quickly things can change. She was highly regarded by the crews with whom she flew because of her thorough operational knowledge and preparation. The passengers had settled into the tight accommodations. Some most likely greeted their neighbors. Others turned to work or reading. Like most young children, Patrick Henry probably watched everything through the window from his seat in row 2.

In seat 6C, Bruce Anglemyer was on his way home to Rocky Mount, North Carolina. Across the aisle to Bruce's left sat Bill Readings, who was returning to Canada after spending some time with his wife, Diane, a professor at Indiana University in Bloomington. A group of friends waited for him in Montreal for a Halloween party. Bill also had a book near acceptance at Harvard, but he did not want to risk jinxing the deal by being congratulated too early.

One row back, Frank Sheridan sat next to Wan Suk Ko.

A hard-core Indiana Pacers basketball fan, Brad Stansberry sat in row 11, probably chatting with his Delco coworker across the aisle about sports.

Barbara Tribble, on her way back to Dallas, occupied seat 13B. She was no doubt eager to tell her husband, Ron, about her son's swearing-in ceremony.

In the cockpit, Orlando and Jeff completed the taxi checklist on the way out to the runup pad. The runup pad holds aircraft out of the way near the runway while it waits for permission to take off. Parking in the runup pad for runway 23R at Indianapolis provides the outgoing crew with a view of the incoming American Eagle ATR-72 from Chicago that lands on 23R. Pilots often judge the landings of arriving flights while they sit at the pad and may have a chance to chuckle as fellow pilots are surprised by a last-second gust of wind or whistle in awe as they "grease" one on.

At about this time, the familiar "door bell" chime in the cockpit would have indicated that one of the flight attendants was calling on the intercom. Sandi would call to inform the crew that their pretakeoff duties were complete, and the cabin was ready for departure.

Captain Aguiar was always eager to help the flight attendants by keeping the passengers informed. Passengers can treat flight attendants harshly, somehow thinking they are linked to the main computers and always expecting them to have the latest information. Orlando would have made an announcement similar to the following:

"Good afternoon, folks. This is your captain speaking. Welcome aboard American Eagle Flight 4184 with service to Chicago O'Hare Airport. We want to apologize for the delay today, but the weather in Chicago is a little worse than it is here in Indianapolis and that has a tendency to slow things down a bit. We are expecting to be airborne at 2:52 P.M. Central Standard Time, that's 3:52 P.M. for those of you still on Indy time. Our flying time to Chicago today is scheduled for 47 minutes at an altitude of 14,000 feet. There are a few bumps out there today, and some rain showers in northern Indiana, but nothing to be concerned about. We have a couple of the best flight attendants at American Eagle working our flight today, and they will do their best to make your flight safe and comfortable. Should you need anything, please don't hesitate to call on Sandi or Amanda. Thank you for flying with us today and, once again, welcome aboard."

On the radio, the American Eagle crew would have heard the United flight to Chicago call for taxi instructions. "Ground, United 501, taxi."

"United Five Zero One, Indy Ground, taxi to runway two three left."

Jonathon Henry, Patrick's brother, and his traveling companion were on the United Boeing 737.

Central Standard Time: 2:40 P.M.

The following radio calls were taped by the various air traffic control facilities.

"Is that fifty-two time gonna hold up for Eagle Flight?" Jeff inquired over ground control frequency.

"Uh, still waiting on the Center right now," came the answer. "I guess not too much going on to Chicago. I'll tell you what, you can just go ahead and return to clearance frequency. They'll let you know as soon as they find out something."

"Okay. Do you mind if we just stay right here on ground with you?"

"You can do that. I can't do nothing else for you, but, uh, you can stay here."

"Yeah, we'll just stay here. You let us know. It's no problem. We're just wondering." Jeff Gagliano was always cordial and accommodating, two unusual traits in an airline pilot dealing with flight delays.

The call from Eagle Flight 4184 to the controller prompted him to call Chicago Flow Control. This position in the Chicago Air Traffic Control Center monitors traffic density and issues and approves the departures of flights.

"Indy request release Eagle Flight 184 to, uh, O'Hare," called the ground controller over the landline telephone.

"EDCT," came the response from the flow controller, meaning "What is his expect departure clearance time?"

"Two zero five two." Times are always referenced as Universal Time, the time at Greenwich, England, based on the 24-hour clock and translating to 2:52 P.M. CST.

"Um, he is released. That fix is in the hold, so he might do some holding when he gets up here, but he's released," came the response from Chicago Flow Control. Eagle Flight 184 had permission to take off and enter the traffic flow toward Chicago O'Hare but could expect to circle somewhere en route until air traffic control was able to fit them in. This would be the crew's second delay. The ground controller called the flight.

"Eagle Flight 184, you can expect a little bit of holding in the air and, uh, you can start them up. Contact the tower when you're ready to go." With this information, the crew could elect to continue to wait on the ground; chances were, however, that this hold would be lifted along the way, and by this time the pilots and passengers were eager to be underway.

"Uh, that's for Eagle Flight 184. Roger," replied Jeff Gagliano.

"Flight attendants, prepare for takeoff, please," came the public address announcement from the cockpit. In the cabin, Sandi and Amanda took a final look up and down the aisle and checked their own lap belts and shoulder harnesses; they had long ago completed their pretakeoff passenger briefings. In the front of the cabin, Sandi Modaff faced backward in her jumpseat.

"Indy tower, Eagle Flight 184 ready for takeoff, two three right."

"Eagle Flight 184, Indy tower, runway two three right, taxi into position and hold," came the response from the tower controller.

Captain Aguiar positioned the ATR-72 on the centerline of the runway as the crew completed the takeoff checklist. "Bleeds and lights to go," Jeff cautioned.

"Okay," came the response from Orlando.

"Eagle Flight 184, runway two three right, turn right heading three two zero. Cleared for takeoff." Eagle 4184 was taking off to the southwest and would enter a right turn to a northwest heading when airborne.

"Cleared for takeoff; right turn to three two zero, Eagle Flight 184."

"Bleeds are off, lights are on. Takeoff checklist is complete," stated the first officer. "Set takeoff power."

With that command, the roles of captain and first officer were modified for this leg of the trip. Jeff became the flying pilot and was responsible for controlling the aircraft. Orlando, while still the captain, was assigned the role of nonflying pilot and was responsible for completing checklists, communicating on the radio, and assisting Jeff.

Both power levers were advanced, and the aircraft began a slow acceleration down the runway. Eagle Flight 184 was heavily loaded and would accelerate slowly and climb less steeply. On the descent, it would pick up speed more rapidly. On the positive side, the airplane would be more stable in the turbulence expected on today's flight as a result of the greater mass. All systems and engine indications were in the green, and, at a predetermined speed, Jeff pulled back slightly on the control column. The nose of the aircraft aimed skyward as the two 2700-horsepower turboprop engines pulled the aircraft into the sky.

A radio message was sent by an electronic device on the aircraft, called ACARS, via relays on the ground, to the AMR central computer system. The message was short. "American Eagle Flight 4184 is off the ground."

Central Standard Time: 2:56 P.M.

"Eagle Flight 184, contact departure."

"One eighty-four, see you," came the reply from the captain now operating the communications radio.

Four minutes after takeoff, Flight 4184 was climbing to 14,000 feet and heading northwest toward Chicago and in radar contact with Indianapolis Center. One minute later, the crew was asked to turn toward the northeast for spacing with other aircraft. Indianapolis instructed the pilots to turn back toward the northwest shortly thereafter, and the flight was handed to radar control at Chicago Center.

"Center, Eagle Flight 184 is with you out of, uh, ten point seven for one four thousand, three ten heading."

"Eagle 184, Chicago Center, is cleared direct Chicago Heights. The altimeter setting at Lafayette is two niner six four."

"Out of ten thousand, the autopilot is on," reported the first officer to the captain.

"Understood," responded Captain Aguiar. "The sterile light is off, but I'm gonna keep them in their seats in this chop. I don't think it's going to get any smoother the farther north we go, but no sense taking any chances."

The "sterile" light is located in the cabin and indicates to the flight attendants that the pilots are not to be disturbed except in an emergency. A flight is considered sterile during ground operations, take-off, landing, and any time the flight is below 10,000 feet. According to FAA regulations and airline policy, no unnecessary conversation or distractions are allowed while a flight is sterile.

In the cabin, Sandi and Amanda were preparing for a beverage service. Even in light chop, airline management insists on everyone getting a snack and something to drink. Some passengers were engaged in conversations. Others were pulling out work, books, or magazines to help pass the time.

In Aurora, Illinois, the location of the Chicago Center, a staff briefing for the night shift was about to begin. South Area Supervisor Michael Debb was among the attendees and was informed of ongoing computer problems, airport flow control procedures in effect, and current and forecast weather conditions. At the bottom of his note sheet, which he posted and distributed to all of his controllers, he wrote: "ICING KILLS—It's your job to know the freezing level in your sector, and the tops and bases [of clouds], . . . That is the fastest way out of the ice. Pass on the PIREPs [pilot reports]. Use departures off your airports to solicit this critical info." When questioned later regarding this nonstandard section of his posting, Debb responded, "Usually, at this segment of my briefing, I try to include some kind of motivational-type saying. Usually, it is from somebody else, but today the emphasis that the meteorologist put on our weather that was affecting our center, I thought that it deserved a mention. At the end of October, we were just in the beginning of the icing season, and I wanted to highlight to the controllers how important it was to stay alert and stay on top of the weather conditions in their particular sectors."

American Eagle Flight 4184 leveled at 14,000 feet and started to be bounced by turbulence; the ride was becoming uncomfortable in some of the building cumulus clouds. Shafts of rain that sounded like slush thrown onto a car by a passing truck pelted the windshield. The crew had experienced worse flights but requested a higher altitude because passenger comfort was a concern. The pilots knew they would not "top" the weather, but they hoped to smooth the flight as much as possible.

"Any chance Eagle Flight, ah, can climb up to one six thousand?"

"Eagle Flight 184, maintain one six thousand." After reaching 16,000 feet, the crew realized the ride was going to be the same no matter what altitude they chose, and they decided to start their descent to 10,000 feet. They were given a crossing restriction of 40 miles southeast of the Chicago Heights VOR, a navigation radio located on the south side of Chicago, at 10,000 feet. At 5½ to 6 miles of progress over the ground every minute, they were closing fast on this limitation. The captain checked in with the Boone Sector of Chicago Center, which is responsible for a geographic area of airspace including northwest Indiana, at 3:13 P.M.

"Chicago Center, Eagle Flight 184 is checking in at one six thousand. We have a discretion, uh, down to one zero thousand, forty southeast of the Heights. We're on our way down now."

During the descent, the flight crew activated the airframe deicing system. Flying north into an intensifying low-pressure area brings lower temperatures, continual rain shafts, ever-increasing and changing winds, and significant vertical instability, causing a choppy and an occasionally rough flight.

Service was completed in the cabin. Occasional gasps would have come from passengers as the aircraft tossed, and everyone experienced G-forces varying from more than one-and-a-half times their body weight pushed into their seats to periodic, short excursions into near-zero gravity.

Passengers often are surprised by the sudden drops experienced in turbulence and, when combined with the unpredictable occurrence of the bumps and the general apprehension during flight in small, commuter aircraft, the stress level of the passengers increases. Experienced flyers tend to hide this discomfort by sinking deeper into their work or reading, while jittery or inexperienced travelers tend to become more aware of their surroundings and investigate the interior and exterior of the airplane with increasing frequency. These individuals sometimes seek solace in the apparent calm demeanor of their fellow riders or the cabin crew. By this time, Sandi and Amanda were stowing the service carts and items in the rear galley area, probably talking quietly. An occasional soft laugh would have calmed the nearby passengers.

Wisps of cloud streaked past the airplane. Rainwater flowed in horizontal rivulets over the portal. The wingtip strobe light flashed in the darkening sky. Passengers could not see the thin molding of clear ice that was beginning to adhere to the wing.

Central Standard Time: 3:18 P.M.

Three minutes after the normal scheduled arrival time of American Eagle Flight 4184 into Chicago, Chicago Center issued holding instructions for an anticipated 12-minute delay. The pilots were directed to fly a racetrack-shaped pattern at the LUCIT intersection, approximately 50 miles southeast of Chicago O'Hare.

The flight conditions were volatile at the time they leveled the aircraft at 10,000 feet. Temperature and atmospheric conditions were again proper for the pilots to deactivate the airframe deicing boots. As the aircraft approached the holding fix at LUCIT, Jeff Gagliano slowed the aircraft to 175 knots (about 200 miles per hour), the maximum airspeed allowed in a holding pattern by the FAA for a turboprop aircraft.

At 3:19 P.M., air traffic control issued the flight an extension to its holding time until 3:45 P.M. Flight 4184 would be one hour late today. Even with the reduced operations at Chicago O'Hare, many passengers on board the aircraft would not make connecting flights. For international passengers, that might have meant an overnight stay in Chicago to catch a flight the next day. Those costs would have been borne by the customers, because American Airlines takes no responsibility for connections that are missed as a result of weather delays.

At 3:24 P.M., Flight 4184 crossed over the LUCIT intersection at 10,000 feet and started a right-hand turn to a reciprocal heading to the airway. The crew reported to the controller that they were entering the hold. At 10,000 feet in the holding pattern, the flight was not sterile as defined by FAA regulation and airline policy. A commercial station played on one of the navigation radios, a common practice among airline crews. The autopilot was flying the airplane, and the pilots monitored its performance and the systems of the aircraft. Until the flight could be released toward Chicago O'Hare, things would be quiet. Power was reduced, and the noise in the cabin settled to a moderate hum.

With the cabin duties completed, Amanda Holberg called the cockpit to see if flight crew wanted anything to drink. Contacting the pilots also allowed Amanda to check on the progress of the flight to ensure they would not divert—one of Sandi's fears on this day. The conversations in the cockpit were recorded by the cockpit voice recorder.

"Now I see what's going on up here," Amanda said, arriving in the cockpit. Jeff turned the music onto the cockpit speakers. "Is that like stereo, radio? You don't have a hard job at all. We're back there slugging with these people."

"Yeah, you are," Orlando said. "We do have it pretty easy. I was telling Jeff, I don't think I'd ever want to do anything else but this."

"What do you all do up here when autopiloting? Just hang out?" asked Amanda.

"You still gotta tell it what to do," Jeff answered.

"If the autopilot didn't work, he'd be one busy little bee right now," said Orlando, and the three crewmembers laughed.

"So does the FO do a lot more work than you do?" asked Amanda.

"Yep." The laughter continued. These pilots got along well, and the atmosphere was cordial, relaxed, and considerate. A crew that gets along on a personal level increases the chances of proper and precise handling of any routine tasks and unusual occurrences.

"Man, this thing gets a high deck angle in these turns," said Captain Aguiar, gently prodding his first officer. As the flying pilot, Jeff Gagliano had control of the aircraft under airline rules. Although the captain has the authority to override any command from his subordinate, Orlando Aguiar tended to direct rather than order. His concern was the slow airspeed in the holding pattern and the weight of the aircraft, which combined to cause an increased pitch angle, that is, nose high, making the ride a little uncomfortable because the aircraft had a tendency to wallow in turns. The captain was hinting that Jeff should put the flaps at 15 degrees, providing additional lift, lowering the nose, and making the ride feel more stable.

"Yeah," Jeff responded.

"We're just wallowing in the air right now."

"You want flaps fifteen?" Jeff asked.

"I'll be ready for that stall procedure here pretty soon," Captain Aguiar continued. "Do you want kick 'em in? It'll bring the nose down."

"Sure." The flaps were placed at 15 degrees. The aircraft autopilot system retrimmed the aircraft, taking the control pressures off by mechanically moving small tabs on the flight control surfaces. The nose of the aircraft came down noticeably, from about 5 degrees nose up to level, and the ride became more comfortable. From experience, the captain had sensed a higher-than-normal pitch angle accompanied by a loss of airspeed in the turns; additionally, having the flaps at 15 gave the aircraft a significant speed protection against a stall.

Ice continued to accumulate on the airframe, causing an increase in drag. A warning chimed, indicating that the electronic sensor located on the underside leading edge of the left wing sensed a change in the harmonic vibration and that the aircraft might have been picking up ice. The crew again selected the proper anti-icing and deicing systems, and what Captain Aguiar unconsciously sensed should have been corrected by the deicing system.

Central Standard Time: 3:38 P.M.

The crew was advised by air traffic control to expect a release out of the hold at 4:00 P.M. The periodic additions of 15 minutes to the flight's delay were not accidental. FAA procedures require a written report be filed if a flight is issued a hold longer than 15 minutes, and controllers dislike paperwork. Unfortunately, these short-interval hold instructions interfere with the pilots' abilities to adequately plan and manage their flight. Jeff and Orlando discussed the new "expect further clearance" time, their fuel status, and notification of the airline dispatcher as to their current status. The pilots spent several minutes discussing the protocols and working on transmitting the required messages on the aircraft communication and reporting system known as ACARS. Amanda returned to the cabin.

By this time, the passengers would have started getting restless and anxious over the delays. Regional airliners are not designed for long-term comfort, and it had been more than two hours since the boarding began in Indianapolis. In addition, the air-conditioning system on the airplane was only functioning at about half level because of a maintenance problem that eliminated the use of one of the aircraft's bleed air sources. The continued turbulence and its physiological and psychological effects probably caused some people to experience queasy stomachs.

At 3:48 P.M., Jeff Gagliano noted the ice accretion on the aircraft for the first time. "I'm showing some ice now," he said. He based this comment on the observation of portions of the wing visible from the cockpit. These wing sections were about 40 feet away from his seat, and on the leading edge of the wings he could see the rubber deicing boots that inflate to break the ice adhering to the wing. The ice should then be swept away by the moving airstream. Inflatable rubber bladders use technology from the 1940s, which was adapted by the plane's designers into a system in which the deicing boots on the ATR operate continuously, more like an anti-icing system. This peculiar adaptation runs counter to the normal experience of propeller aircraft pilots and has been held indirectly responsible for some previous ice-induced control problem incidents and one accident, as documented in letters from the Air Line Pilots Association (ALPA) and reports by the FAA and the National Transportation Safety Board (NTSB).

In the dark conditions, neither pilot saw the slight amount of ice adhering to the wing behind the limit of the boots. The pilots discussed the nicer ride with flaps down, and Orlando Aguiar warned, "I'm sure that once they let us out of the hold and (we) forget they're down, we'll get the overspeed."

American Eagle Flight 4184 was operating in a narrow performance envelope with the ice accumulating on the aircraft. FAA regulations limited the flight to an airspeed no higher 175 knots in the holding pattern. To remain within the horizontal limits of the protected area allowed in the congested airspace around Chicago for this maneuver, the crew was required to turn the aircraft at a standard rate of 3 degrees per second, operating in a "high bank" mode. With the weight of the airliner at slightly over 44,000 pounds, the minimum speed allowed for this degree of turn was 163 knots. The turbulence existing at the time caused routine 10-knot fluctuations in the indicated airspeed. Every FAA regulation is breakable in an emergency or when, in the pilot in command's judgment, safety dictates; however, in this case, neither pilot realized a dangerous situation was developing. Everything in their conversation and demeanor, as recorded on the cockpit voice recorder, indicated a calm, routine situation. The captain went to the cabin to use the bathroom and returned several minutes later.

Central Standard Time: 3:55 P.M.

"We still got ice," Jeff said. The flight crew then resumed their conversation regarding the release time from the hold, communications with the airline dispatcher, and connecting gate information. Captain Aguiar called the airline's operations at Chicago O'Hare on the radio, leaving his first officer to deal with air traffic control.

Central Standard Time indicated by the flight recorder: 3:56:14.

"Eagle Flight 184, descend and maintain eight thousand," the controller directed. Jeff apparently did not hear this transmission, and the call went unanswered until 3:56:27 when the controller reissued the directive.

3:56:31. "Down to eight thousand. Eagle Flight 184."

3:56:44. "Eagle Flight 184, uh, should be about 10 minutes, uh, till you're cleared in," came the news from air traffic control.

3:56:50. "Thank you." This was the last recorded transmission identified from the flight to air traffic control. The captain's conversation with operations continued, and the first officer commented to him over the intercom, "They say 10 more minutes." This was apparently unheard, because as Orlando completed his conversation with operations he queried Jeff. The flight reached the southern end of its outbound holding leg, and Jeff commanded the autopilot to begin a right turn to the northwest, at a reduced bank angle because of course drift, and started a descent to the newly assigned altitude. The aircraft was traveling at 175 knots. The overspeed for the flaps at 15 degrees is officially 185 knots, but the warnings on the ATR routinely sound at about 180 knots.

Some passengers sensed the beginning of the descent from the slight nose-down attitude and the reduction of the engine power. A few felt the beginning of the turn, although in the clouds the sensations were deceiving.

3:57:16. "Are we out of the hold?" asked Orlando.

3:57:17. "Uh, no, we're just going to eight thousand," answered Jeff.

3:57:19. "OK."

3:57:20. "And, uh, 10 more minutes she said. . . . Oops." The overspeed warning beeped. The aircraft was heavy, and the use of normal power levels for a descent caused an acceleration. The overspeed warning sounded at 181 knots. The airliner passed through an altitude of 9600 feet.

3:57:26. "We, I knew we'd do that," Orlando commented.

3:57:27. "I'm trying to keep it at 180," said Jeff, referring to the maximum speed at which the overspeed warning would not beep. He retarded the throttles to flight idle; almost simultaneously one of the pilots began the flap retraction by selecting the flap handle to the zero setting. The autopilot began trimming the aircraft to the changing configuration as the airspeed dropped from 186 to 182 knots during the minor changes in the pitch axis.

3:57:29. A repetitive thudding sound was picked up by the intercom. The sound was heard in the cabin. Captain Aguiar reacted with a surprised "Oh." The pitch trim worked almost continuously for about three seconds as announced by a "woo woo woo" noise from a device known as the "whooler."

3:57:32. "Oops, damn!" Jeff yelled. Thudding sounds and aircraft warnings were joined by a rattling noise, and about a half second later as the flaps reached the fully retracted position, the autopilot disengaged and warned the pilots with repetitive, rapid, triple chirps. The aircraft passed 9130 feet. Passengers would have become acutely aware of the external sounds.

3:57:33. "Shit," Jeff exclaimed, probably grabbing for the control column. When the autopilot disengaged about three-tenths of a second earlier, the wheel snapped to a full right-wing-down position. The aircraft rolled violently to the right, stopping at a 77-degree, right-wing-down position. Everyone aboard the aircraft was tossed to the left and then quickly to the right as the G-forces caught up.

As he grabbed the wheel, Jeff must have activated the transmit switch for his radio. An unintelligible, short transmission was recorded in the Chicago Air Route Traffic Control Center at 3:57:33. Near Wausau, Wisconsin, Jeff's best friend, Tony Zarinnia, was landing his Beechcraft airliner. He noted the time as halfway between 3:57 and 3:58 when he clearly heard Jeff's voice yelling "help" in his headphones.

As the flight data recorder later showed, Jeff attempted to force the control wheel to the left to counter the roll, and the wheel unexpectedly snapped to the left, catching him off guard. Neither pilot could have known what was happening to the aircraft. However, both must have known that a frightening control anomaly gripped the aircraft. The ATR was accelerating rapidly.

An intermittent heavy irregular breathing was heard on the cockpit microphone. Eyes desperately scanned the instrument panel for clues to the plane's attitude. The flight instruments were the only way to determine the aircraft's position; yet to the crew's horror, the instruments were a jumbled mess of confusing and terrifying information. Because the inner ear is fooled by the rapid head movements caused by the aircraft accelerations and gyrations, the equilibrium of the pilots and that of the passengers and flight attendants in the cabin was affected. Sensations of orientation rapidly became unreliable, even dangerous. As they were trained, the pilots were most likely thinking, "Concentrate on your flight instruments. Trust your instruments." However, logic must have told them that this could not be happening. Nothing made sense. To reverse the last action before the upset, which is standard pilot procedure, Orlando reset the flap handle to the 15-degree position, all the time knowing the computer would not allow the flaps to extend.

3:57:36. "OK?" Orlando asked to ensure Jeff was still in control and responding. The airliner began to recover to the left. The bank angle reduced to about 60 degrees right-wing-down, but the nose of the airplane was 15 degrees below the horizon, and the nose-down pitch steepened. The altitude was 8854 feet. G-forces were increasing, and all aboard were pushed into their seats at one and one-quarter times their body weight. In the cabin, people grabbed seat armrests.

3:57:39. "Oh, shit!" Jeff was battered by the controls and realized that the aircraft was not responding. The aircraft fought his attempts to restrain its erratic movements. The rapid thudding sounded like the beating of the blades of a large helicopter. The plane again rolled to the right, rapidly rotating at more than 50 degrees of bank per second, faster than some acrobatic aircraft. The aircraft was on its back, then right side up, and again on its back. G-forces exceeded two and one-half times normal.

3:57:42. "OK!" Orlando saw Jeff fighting the aircraft and understood that this was not the time to take control. His first officer had been wrestling the aircraft and probably had a better feel for it. The aircraft plummeted with its nose down 55 degrees; it passed through 7940 feet at 213 knots. Wind noise increased. Throttles were fully retarded, but the airspeed continued to increase rapidly. A growling sound began. Jeff would have had to fight with every ounce of strength. With the unpowered flight controls, he was faced with a rapidly changing and high resistance to his efforts to control the aircraft. He attempted to use the rudder to help right the ship but was restricted by another design modification that limited the rudder travel to 2.5 degrees at more than 185 knots, a speed that the aircraft had long ago exceeded.

3:57:44. "Alright man. . . .," Orlando said.

3:57:45. "OK, mellow it out," he said. The aircraft was exceeding its maximum operating speed of 250 knots. Flight 4184 was falling at 30,000 feet per minute. The sensations in the aircraft elicited terror in the cabin. The pilots remained calm and continued to try to regain control.

3:57:46. "OK." Jeff was still trying to regain control. His flight instruments presented him with a confusing scenario. His attitude indicator showed a solid brown circle representing the ground in the severe nose-down position. The altimeter unwound wildly and made one complete rotation, indicating the loss of 1000 feet, every two seconds. The airspeed indicator climbed further past the maximum speed. The aircraft structure was overstressed by the efforts at recovery. The wind roared.

3:57:47. "Mellow it out." Orlando tried to keep Jeff from fixating his attention and freezing up.

3:57:47. "OK." Jeff worked to regain control, but the airliner continued its nose-down pitch to 66 degrees in a near-inverted dive.

3:57:48. "Autopilot's disengaged."

3:57:49. "OK."

3:57:52. "Nice and easy." The aircraft passed 3400 feet, still in the clouds and descending at more than 36,000 feet per minute at more than 350 knots indicated airspeed. The aircraft dropped at 600 feet every second; most passengers were frozen in fear.

3:57:54. "Terrain, whoop, whoop!" The ground proximity warning system sensed the approaching earth and warned the crew.

3:57:56. "Aw, shit." The aircraft broke out of the clouds, and Jeff saw the ground for a second. He pulled back on the control column in an attempt to raise the nose. The aircraft exceeded 373 knots. G-forces reached 5.2, and the aircraft broke. The outer 10 feet of both wings and the horizontal tail separated from the airliner. The aircraft smashed the ground and disintegrated into pieces.

The cockpit area microphone picked up a loud crunching sound, and the cockpit voice recorder recorded the time of impact as 3:57:56.7 P.M. At 3:57:57.1 P.M., the recording ended as power was lost.

Sixty-eight were dead, their bodies shredded and strewn over 8 acres of Indiana farmland. Within seconds, the quiet returned, and the wind and rain were audible again. Rustling and blowing papers, important an instant earlier, were now debris. Only 36.7 seconds had passed since the first officer began the descent to 8000 feet.

4

Response, reflection, and remorse

"It's over."

Seventeen-year-old Brad Anglemyer uttered these words while he watched the family's television. The words came in response to his mother's hopeful statement, "Dad is okay." Sandy Anglemyer was attempting to alleviate the fears of her younger son and also calm her own anxiety over the news story that repeatedly flashed into her home in Rocky Mount, North Carolina. Although the announcer originally reported the possibility of survivors of the airline crash in northwest Indiana, the next update dashed all hope.

No one survived the crash of American Eagle Flight 4184 bound for Chicago from Indianapolis; the rescue workers had called off the search. On the screen, the picture that played on CNN throughout the night showed only darkness broken by flashing red and blue lights of emergency vehicles that were not needed.

Sandy's carefully planned and prepared turkey dinner was forgotten. Had it been only two hours since she had spoken with Bruce before he boarded the flight in Indianapolis? Or was it three? Sandy remembered Bruce had complained of not feeling well, and the only remaining possibility in her mind was that her husband of 25 years had missed the flight. Grasping at straws, she repeatedly dialed his cellular phone number.

As she listened to the distant rings, praying for the answer of a familiar voice, Sandy noticed her son nervously pacing in the far corner of the room. Brad was the captain of his high school swim team and an honor student, and he had a full scholarship to North Carolina

State University the next fall. His brother Jon, 21, was in college, an hour's drive from Rocky Mount.

"If the worst has happened, we cannot tell Jon over the phone," Sandy remembered thinking. "We have to drive to school and tell him in person."

Jon Anglemyer discovered at 4:00 P.M. that no Halloween party was planned at his fraternity house, so he organized one at the last minute. As the party was getting rolling, Sandy and Brad were leaving their home and beginning the tortuous drive to deliver the news.

When Sandy and Brad entered the fraternity house and informed Jon of the accident and their worst fears, a hush fell over the party. Bruce had been there with Jon on a number of occasions, and many in attendance had met his personable father. Bruce Anglemyer understood the meaning of camaraderie and always cherished his association with his own Delta Sigma Pi brothers more than a quarter of a century before. These college men had a saying—"The world will be a better place when a Delta Sigma Pi passes by"—and from the looks on the faces of Jon's brothers, it was understood that nobody personified that saying as much as Bruce Anglemyer. The pain of the three family members was felt by the others in attendance who watched in silence, and it was because of these people that the family would have no shortage of friends to comfort and assist them in their upcoming challenges.

In South Bend, Indiana, 500 miles away, the home where Bruce had spent the last two nights of his life was quiet except for the television watched by Arthur Bauernfeind. His wife, Mary, was unaware of the content of the news flashes that interrupted his early-evening programs. The first news flash announced a plane crash in northwest Indiana; the next contained flight information, and it became increasingly clear to Bruce Anglemyer's 88-year-old stepfather that his wife was in for terrible news. Arthur kept the secret from Mary that evening because he did not want her to be alarmed unnecessarily, and if it were true, nothing could change anyway.

At 4:00 P.M. on Halloween, Tom Severin left the office in a northern suburb of Chicago that he shared with Patricia Henry, his sister-in-law and business partner. As he stepped into the cold rain falling from the gray afternoon sky and the gusty, northeast wind, the normally reserved private investigator raged at the sky. For reasons he did not then understand, he erupted with a sudden, primal scream. Almost embarrassed by the outburst, Tom pulled his coat tightly and dashed

through the weather to his car. He did not yet know about the accident that had claimed Patty and Patrick two minutes earlier.

In some Native American folklore, the coyote is a trickster and a foreboding omen, but when one ran across the path of Barbara Tribble, her son Garvin Senn, and his wife Carla, as they were driving north to Indianapolis on the evening before the swearing-in ceremony, nobody thought much about it. The itinerary for Barbara's travel was a toss-up: fly into Evansville, Indiana, where Garvin and Carla lived and then back again after the ceremony, or fly into Indianapolis and avoid the four-hour drive back to Evansville after her son was sworn in. Since she was there for the weekend and wanted to get back by Monday night, the choice was simple. Besides, the fare had been cheaper with a connection in Chicago to Indianapolis.

It was after 8:00 P.M. Chicago time when Garvin and Carla arrived in Evansville. As he settled in to watch the Bears-Packers game, Carla telephoned her mother to relate details on the ceremony and the weekend visit with Barbara. She was surprised by the urgency of her mother's questions regarding the timing of the flights Garvin's mother was on back to Dallas. Carla explained that they had arrived at the Indianapolis airport just in time for Barbara to catch her airplane, which was scheduled to depart shortly after 3:00. The airline personnel had told them the flight was still operating on time; Carla knew the trip was about an hour. Then her mother told her that an American Eagle commuter flight had crashed between Indianapolis and Chicago at about 5:00 P.M. Indianapolis time. Carla's blood turned cold. She knew Barbara had been flying American Eagle, and she had talked about flying on a small commuter plane. But it must have been a later flight that crashed. Carla had to tell Garvin about the accident.

"My mother says there has been a crash of an American Eagle airplane between Indianapolis and Chicago around 5 o'clock," she said. "That's too late to have been your mother's flight, isn't it? Why don't you call Ron?"

"Turn on CNN and I'll call," her husband said while thinking to himself about the time of the flight and how he would most likely find his mother safe in Dallas.

"Hello," came Ron Tribble's quiet voice.

"Ron, it's Trey," said Garvin. "Is Mom there?"

"No, she's not," answered Ron. "I went to the airport to pick her up and waited for two or three flights from Chicago to arrive, and she wasn't on any of them. So I came back home and figured she would call when she finally did get back."

"There has been a plane crash, Ron. CNN is saying it was from Indianapolis to O'Hare. Wait a minute, there's a phone number on the screen to call for information. I'm calling the airline so your phone can stay open in case Mom calls."

Garvin's experience of telephoning the 800 number reported by all the news services was repeated throughout the night by frantic people in search of information on the fate of their relatives and friends. Some people had worse encounters on the telephone: some left their phone numbers and relative's name with a request to be called as soon as information was available and never heard from the airline. After several attempts, the phone was answered automatically, and Garvin's call was put on hold for 20 minutes.

As the electronic buzzing of the telephone on hold began to annoy and frustrate him, Garvin began to fear the worst. His mother had always been the informational hub of the family, and he could always call her to fill him in on what was new. His sister, Melissa Nash, was geographically closer than Garvin was to his mother, and the two women had finally become close after years of tension. His mother's joy had overflowed during and after his swearing-in ceremony, and when the photographs were developed, they revealed her happiness and pride.

Finally a human voice came onto the line, and Garvin attempted to ferret out information. To his frustration, few details were available that could be shared by the airline with an almost-anonymous caller, and the best he could do was leave the family data and his mother's name.

"Can you at least tell me the flight number?" he asked. "We need something to go on."

"It was American Eagle Flight 4184 from Indianapolis to Chicago."

What Trey had not realized was that during his call, the media had learned the identity of the missing flight. The information was no longer secret. With the number of the doomed airline flight, Garvin dialed his stepfather with trepidation.

"Ron, it's Trey. Do you know the flight Mom was supposed to be on today?"

"I have a copy of her itinerary here from the travel agent." Garvin heard the sound of rustling paper as Ron sorted out the flight information. "It says here AA 4184, must be flight 4184 between IND and ORD."

"Oh, my God. That's the flight that crashed. I think that was Mom's flight."

The conflicting reports on television kept alive the hope for survivors. After all, crashes almost always have survivors; even airplanes smashed to the ground by thunderstorms have survivors. Maybe the airplane had only made a forced landing; after all, it was a commuter airplane with propellers. Being the cynical new attorney, Garvin would not trust the accuracy of the press reports, which could be highly erroneous in the early hours of a disaster. That same cynicism served him and his family well when dealing with the airline propaganda in the days to come.

Somebody had to locate and inform Melissa before she heard the story from the newscasts. A call to her home revealed the young mother had been at school most of the day with her six-year-old son, Andrew, and she was out trick-or-treating with Andrew and three-year-old Avery. The three were with another family, and after several phone calls, the phone rang in the car of Melissa's friend at about 9:30 P.M. Dallas time.

"You need to take Melissa home and come home yourself," the friend's husband said. "There's a problem." The tone of the message was ambiguous, and the women joked that maybe the friend's husband was leaving her. As she dropped Melissa and her boys at their home, her friend had no idea as to the depth of the troubles.

When Melissa walked in the door, she found people in her apartment; her husband looked ill and wanted to talk to her in the bedroom.

"Trey called," he started. "They can't find your mother's plane."

The impact of the message flooded Melissa's mind with thoughts, and the sounds of the outside world did not penetrate the numbness that surrounded and protected her. She had wanted to attend Trey's ceremony with her mother. How would she cope without her best friend? What would have happened to her two young sons if she had made the trip? Was her mother all right? Was she hurt or trapped or frightened? There were many questions but no solid information. Several hours later, the Tribble, Senn, and Nash households received the first painful answers.

Between 2:00 and 3:00 A.M., American Airlines called Ron Tribble, confirming that Barbara had boarded American Eagle Flight 4184 and confirming the reports of no survivors. With his heart heavier than it had ever been, Ron Tribble called his stepson and stepdaughter.

In the northern Indiana darkness, the wind and rain continued while county workers hauled truckloads of gravel for earthmoving machines to create a road into the muddy soybean field. State police worked with county and local police to coordinate the security of the area from a command post. Lights from vehicles moved along the gravel roads, but nobody trespassed into the secluded 40-acre parcel containing the remains of Flight 4184. The Power Ranger action figure Barbara Tribble found for her grandson Andrew during her Sunday shopping excursion in Evansville lay in the wet debris.

Trick-or-treaters were calling on the Severin household in Glenview, Illinois, when Jim Henry came to see one of his daughters at about 4:30 P.M. Terri Severin greeted her father and immediately saw that the color was drained from his face. He told her a plane had crashed between Indianapolis and Chicago. He wanted Terri to call the airport to find her sister and nephews, Patricia, Jonathon, and Patrick Henry. He seemed to know they were on the airplane that crashed, but Terri was unfazed. Something like that could never happen to her family, and she nonchalantly found the number for the Indianapolis airport and paged her sister. As only silence answered her plea, Terri realized that her father was terribly worried. He looked sick. He knew.

Finally, a glimmer of hope arose as someone told Terri that a Pat Henry was being sought by another party. Although her sister went by Patricia or Patty, and nobody ever referred to Patrick as Pat, Terri pursued the lead. Ironically, a Pat Henry was at the airport acting as an early spokesman for American Eagle, and as the time approached 7:00 P.M., Terri focused on the fact that Patty had held tickets on United Airlines, not American Eagle. Airlines, however, do not release passenger information, especially over the telephone, and the nervous sister could not find out that the group split up and traveled on separate planes or that Jonathon and Patty's friend had arrived on the United flight and were currently waiting at Chicago O'Hare. No information was available about the status of Flight 4184, even though the agents might have wanted to help, because on the notification of the accident, the flight's data was secured in American Airline's computer system to thwart curiosity seekers and avoid the inadvertent release of information to the public or news media. The best intentions can cause families to suffer needlessly or to have joy dashed if erroneous information is provided. Individuals may travel under another name or use someone else's ticket, so exactly who was aboard the flight was not as simple as pulling up a name on the computer screen.

In the interest of accuracy, the same cold, heart-wrenching process that has been criticized by the families of victims for decades had begun at many airports across the country where relatives and friends gathered to wait for loved ones who would not arrive. They were hustled into small rooms with no contact or information for hours while officials sorted out details on the manifest of the flight. At some airports, family members were separated from other people waiting for arrivals, sitting and waiting and jumping in apprehension at every opening of the door.

Things were different when Jim and Ruth Henry left for Chicago O'Hare in search of answers around 8:00 P.M. that night. After speaking with several people, they were escorted into a room already occupied by a number of people clustered in small groups. Eyes searched the identities of every newcomer, yet there was little interaction among the different camps as each suffered quietly. It was about the same time that Patty's friend, now frustrated and concerned, and Jonathon left Chicago O'Hare for home. After several hours of waiting, during which each minute seemed like an eternity, the answer the Henrys dreaded came— Patricia and Patrick had boarded the doomed flight. The news was contrasted by the joy in Jonathon's safe arrival.

The direct flight from Indianapolis to Washington, D.C., was quicker than connecting through Chicago, and Jeffrey Lietzan was waiting with a big hug and kiss for his fiancée, Alison Smith-Field, as the stream of passengers exited the jetway and spilled into the terminal. Alison knew he was waiting for her, and he was certain she could not miss him as he strained to see her. The flow of people surged, then slowed to a trickle, and finally stopped. As the area cleared and the agent moved to close the jetway door, Jeff realized that Alison had missed her connection. He walked up to the desk where the agent was changing the posting of flight numbers and the outbound destinations.

"Excuse me. My fiancée was supposed to be on that flight from Chicago, and I was wondering when the next flight will arrive or if she maybe got delayed?"

"I'm sorry, sir," replied the agent. "I cannot give flight information on our passengers, but the next flight from Chicago will arrive in less than two hours."

"Well, can I find out if she is on that flight? She was on American Eagle from Indianapolis to Chicago and then here. I would just like to know so I can plan."

The agent had heard of the accident flight and knew it was coming from Indianapolis. However, she could not tell the young man anything no matter how much she wanted to help. Possibly, to lead the man in the direction that could provide an answer, she said "Maybe you could call somebody."

Jeff took the advice and called his father, not knowing that Alison's brother Greg had already heard about the crash and was heading back out to the airport in Indianapolis where he had seen them off. Jeff found a phone and called his father and complained, "They won't tell me anything." He was concerned and slightly irritated. Jeff's father contacted Doug Smith, Alison's father, and the two discussed the situation. Doug had already heard about the accident and knew Greg was on his way to the airport to find out details. He told Jeff's father, "It doesn't look good; looks like trouble." The two men agreed that Doug would tell Jeff.

Between calls, Doug got word from Greg at the airport in Indianapolis that there were no survivors of American Eagle Flight 4184 and broke the news to Jeff.

Carole Modaff heard about the accident when the news flash interrupted the regular program on the television in her living room around 5:00 P.M. Not many details were known at the time, and even though she did not know her daughter's schedule, intuition told her it was Sandi's flight. Carole broke out in a cold sweat and had to sit down while her mind searched for any details of her conversation this morning that could have proven the foreboding feelings wrong. The next announcement cemented her fears; it was confirmed the lost airliner belonged to American Eagle and that it was believed to be an ATR-72. Carole knew Sandi loved to fly the ATR-72; it just had to be Sandi.

Carole had two other daughters who were flight attendants for American Eagle. Missy was at home this Halloween. Kim had married and left her job as an American Eagle flight attendant, so Carole believed that, if anybody would understand what was going on, it would be Missy. Missy did not know more than anyone else during the early hours after the accident, but she did have some insight into how Sandi felt about the weather because her sister had called from Indianapolis. Sandi had told her sister she hoped the flight back to Chicago O'Hare would be canceled because the weather was so windy and rainy, and she was scared. "Scared" was not a word Sandi used lightly; she loved to fly, and as an experienced flight attendant, she had flown in rough weather. Flight 4184 had originally been delayed, but as Sandi was on the phone with her sister, an American Airlines gate agent found her and said the airplane was going. For the last time they said their goodbyes.

Sandi's special dress would never be worn, and her bright future with her premonition that she would soon marry would never happen. A senior ATR captain at the airline mourned in his own special and private way because he loved Sandi and would never get to see her in the dress she had packed so carefully for their reunion dinner that Thursday.

Little doubt remained as to the identity of the flight that had crashed. The growing group of news media in the terminal were thrusting cameras into the faces of people in pain. *USA Today* published a photograph the following morning of an anguished Pat Sullivan, whose brother was lost on the flight.

The details spread slowly from Indianapolis to the families scattered across the globe. Many relatives were contacted in the middle of the night, and those who were aware of the tragedy waited for the phone to ring, hoping for the best, only to have their worlds shattered by the news in the early morning hours.

Sharon Holberg had planned to fly to Chicago to visit her daughter during the coming weekend to share the new experiences, such as riding the elevated train and visiting the Sear's Tower, that Amanda had so excitedly described during their telephone calls the previous week. Mandy, as she was called by her family, had recently completed flight-attendant training with American Eagle and was stationed at Chicago O'Hare. Chicago was far from her home in Houston, but at 23, Amanda wanted to see the world. On the morning of October 31, she flew her first trip as a flight attendant, and she had reported to the crew coordinator early, as all new hires do. Taking her under his wing, the crew coordinator introduced her to the captain and senior flight attendant for her trip. Now Sharon Holberg sat by the telephone where she had heard the news.

Amanda had been bitten by the flying bug early. At age four, she told her mother she wanted to be a flight attendant. She also wanted to be a cheerleader; however, being only four, she had time to make up her mind. By the time she had gone to college, Amanda had set her sights on a career as a television news anchorperson, and her mother had thought Mandy's dream of becoming a flight attendant was forgotten. Southwest Texas State University in San Marcos had a television station on campus, and before long Amanda was working on the air. By the time she graduated in May of 1994, she had experience on both sides of the camera, and her communications professors tried to guide her into seeking a small market to launch what could have been a stellar career. Before she would settle down into a career in television, Amanda answered the call of her earliest dreams—to fly and travel. "She got an offer from American Eagle," remembered Sharon, "and although I was shocked at how low the pay was, I had to support her in this because I knew she would never get on with her life until she got this out of her system."

"Scheduling," came the curt response at the Eagle Operations Center in Dallas.

"Hi, this is Jeff Collins, ATR captain based at O'Hare."

"Yeah, Jeff, what can I do for you?"

"I saw the news on the television about one of our flights, and I was calling to find out some details and the identity of the crewmembers."

"Not much I can tell you, detail-wise, but it looks like no survivors on 4184 from Indy to O'Hare; ATR-72, I believe."

"And the crew?"

"What I have right now is officially 'unofficial' . . ."

"Yeah, I understand."

"Captain Orlando Aguiar, First Officer Jeffrey Gagliano, flight attendants Amanda Holberg and Sandi Modaff."

"No!" Jeff Collins threw the telephone across the room, and it split in half as it struck the wall. Jeff wondered how to break the news to his wife, Sandi's sister.

"Jeff, are you alright?" Kim called from the other room. "Who was on the phone?"

"I'm going outside to get some air," Jeff answered, and without a glance at the pieces of the telephone scattered on the floor, he poured a stiff drink and walked into the darkness to be alone with his thoughts.

Lewise Morris's daughter Charlew remembered her mother as looking exceptionally lovely the day they traveled to the airport for her return to Florida after a weekend wedding. Shock surrounded the Morris family two hours later when the news of the accident in Newton County reached them, and they realized they had to break the news to their father, who had remained at home because of ill health. After the family told Charles Morris about the sudden death of his wife, he remarked at the irony: "We were married in Newton County, Indiana."

Orlando Aguiar made a call to his wife, Dawna, from Indianapolis and talked about the upcoming christening of his daughter, Marisa. He told her he had spoken with the chief pilot about needing some time off around November 6 for the event, but he was wondering about possibly postponing it so he could fly and earn some extra money to help prepare for the new baby they were expecting. He picked up trips on Halloween, and each day he worked would help their finances before the baby arrived.

"I'm playing cards with the new galley wench," he joked to Dawna, a former flight attendant. She knew the vernacular of pilots and smiled. It was one of the last things he said to her.

The Chicago Bears' football game was on television when the telephone rang at the home of John and Kay Thompson, and John answered. It was a friend of his daughter, Dana. The friend told John that Dana was to meet some of them to attend the football game after her return from Indianapolis, which should have been hours earlier. When the friend mentioned the plane crash, John's knees almost buckled from the sudden rush of fear. Just then, the telephone number for American Airlines flashed on the screen, and John jotted it down.

Although the airline set up a telephone line for families and friends to call to receive information, information was the last thing flowing from the company during the evening hours. The airline had a responsibility to ensure identities were correct, and the telephone line served as a method of collecting names for callback. The flood of calls must have been heavy, because it took many families hours to get through, and they were frequently put on hold for long periods. John Thompson got through after numerous busy signals, and when he finally did reach someone, there was no information or manifest for the flight available. He left his name and telephone number.

Startled by the ringing telephone at about 10:45 P.M., John took a deep breath and looked to his wife for strength before answering. It was Dana's boss from the Fox Network saying he was concerned about her, and he confirmed that nobody had heard from her since she left her last client's office in Indianapolis. Four hours later, in the middle of a sleepless night, the telephone again rang, and this time John knew it would be bad news. It was American Airlines, and the anonymous, sympathetic voice informed him they had confirmed Dana was aboard the flight and that there were no survivors.

For those family members who remained at home, Halloween night was filled with the ironic combination of grief and gleeful trick-or-treaters. Friends and family gathered at the home of Cheryl Dwyer to console her husband, Dennis, and remember the person that meant so much to them all. Cheryl had not been there for the trick-or-treaters after all, and Dennis had not been able to cope with them after learning of the accident. As people came into the home, they noticed the heartbreaking sight of the unopened bags of candy lying on the table in the hallway.

In African culture, there are no losses; many Africans believe life follows death, but when American Eagle Flight 4184 crashed, it caused a large loss to the South African and American cultures. Elkin Thamsanqa Sithole was born in South Africa on April 14, 1931, and nearly from the beginning he was setting marks for his Zulu people. He was the first of his small community of Blaauwbosch in Newcastle to go to college and to earn both master's and doctoral degrees, the first to take an airplane trip, the first to leave the village and go overseas, and when the first all-race election in South Africa was held, he was the first person in line at the South African consulate to vote. Finally, to the grief of all freedom-loving people around the world for whom he struggled tirelessly, Dr. Elkin Sithole became the first black person from his small community to lose his life in an airplane crash.

Dr. Sithole loved to teach, to hand down the bedrock of African culture to new generations of students, and especially to ignite the flame of desire for knowledge. Dr. Sithole's field was ethnomusicology, the study of the ethnic heritage of music, and it was not unusual for passersby to see the professor demonstrating a folk dance or hear his golden baritone voice reverberating to demonstrate a particular point. He had the ability to motivate his students to learn and take pride in their heritage, and it is not surprising he was named professor of the year on a number of occasions. Sharing pride in who we are and where we come from, regardless of race or heritage, was one of his favorite pursuits, and it was attendance at a conference on ethnomusicology that placed him aboard American Eagle Flight 4184. One of his greatest desires was to return to his native South Africa, and two weeks before the accident he related his plans to retire in 1996 and return to his homeland, not to stop working, but to pursue the reorganization of the university music departments and rewrite the country's national anthem.

When the silence returned to the field where the flight came to rest, Elkin Sithole's voice survived in the hearts and minds of the thousands of people he had taught to take pride in themselves. Dr. Sithole's middle name, Thamsanqa, in English translates to "blessing," and if ever someone personified the meaning of his or her name, it was this man.

Many businesses were affected by the sudden loss of employees and friends, but none more than Delco Electronics, which lost four employees. Delco, a division of one of the largest corporations in the world, showed it had a much greater heart than did the airline for its employees. The treatment of the pilots' families by AMR Corporation, parent company of American Airlines, made the Gaglianos and Aguiars feel as if the crew had caused the accident just to irritate their employer. Delco could have been cautious (it had a liability because the employees were on company business when the accident occurred), but the goodwill between the organization and the families helped ease the pain and anger.

Delco issued orders that anything the families needed was to be provided; if they needed someone to help with the children, if they needed a ride, or if they needed to have someone do the dishes, it would be done. They would do it without a discernible thought for recognition or ulterior motive; Delco proved that even a huge corporation can have a heart.

In contrast, according to Captain Aguiar's widow, Dawna, representatives of Simmons Airlines entered her husband's car, which was parked in the employee lot at the airport, and removed personal items that were never returned to her. Anne McNamara, senior vice-president and chief counsel of AMR Corporation, later told me that Simmons acted under the direction of the FBI. Jeff Gagliano's parents, Al and Cindy, reported that they heard rumors of management's comments about the crew's actions. Simmons employees never heard of any apologies being made to the families.

A pilot overheard Ed Harvey, director of flying for Simmons, on the telephone discussing another crash of an ATR-72 in Taiwan a couple of months later. Ed reportedly said, "Jesus, that's all we need is for that one to be blamed on ice." And then he laughed.

Cindy Gagliano knew her husband, Al, would not be home early on Halloween night because he was helping with a fundraiser for an association of restaurant chefs. By evening, activities at the ranch were on schedule even though Halloween was one of the busiest days of the year because of the haunted hay ride.

"I remember the big screen television being on with Monday Night Football when I came into the restaurant with some supplies for the event," Al recalled. "And then there was a news flash about the crash of an American Eagle ATR-72 from Indianapolis, and I knew. I said, 'That's my son.' And everybody looked at me thinking that it couldn't be, but I repeated, 'That's my son, I've got to go.' The phone was busy for two hours at the ranch, and I drove home; that is when the operator broke in with the emergency call from the airline telling us of the accident."

When you cut away all of the peripheral stories, the telephone calls to the homes of relatives and friends throughout the world summarized the accident, but it was the human issues that expressed the depth of tragedy.

Two families were spared the sorrow. Maria Peterson decided to rent a car in Indianapolis and drive to Chicago with her two-year-old son, Christopher. The pair had been in Logansport, Indiana, over the weekend for a wedding, and apparently Maria did not feel comfortable with the weather and the flight to Chicago O'Hare on the small airplane. She still held the unused tickets for that leg of her travel back to her home in Santa Barbara, California. Their replacements on the airplane were another mother and young son, Patricia and Patrick Henry.

The other family initially suffered through the night because of the travel plans of one of their relatives. A son was on his way to visit his mother for her birthday party and missed the flight. Hundreds of miles away, the announcement of the flight information caused pain and grief, while near Indianapolis, the son watched Monday Night Football and fell asleep. He called the next day.

At about 9:00 P.M. on Halloween night, I was watching television in my motel room in Cedar Rapids, Iowa, for any news of the accident. One of our flight attendants called me to say she had spoken with a classmate, and the names of the crew were being discussed. She had a question on the identity of the pilots because she did not recognize the names, but she knew she had the correct names of the flight attendants. I was shocked by the news, and that knowledge deepened my despair. I called my wife, Linde, with the news, and she cried. She had known Sandi. Fifteen minutes later, Steve Henry, my captain, knocked at my door. Steve had spoken with Sue Larson, a senior flight attendant who was manning the phones. The identity of the crew was confirmed, and he spoke the names in a quiet tone.

"The crew was Orlando, Jeff Gagliano, Sandi Modaff, and a new flight attendant, Amanda Holberg. She just started this month." "Jesus," I murmured. I could feel myself pressing my lips together the way my wife says I always do when something is wrong. "I was hired with Jeff."

"Amanda was on her first flight out of IOE [initial operating experience]," he added.

"It was her first day? Any word on the suspected cause?"

"Nothing yet. Everybody is headed to Indiana to start working on it in the morning, but they sound really concerned."

I went back into my room and thought of my friends, adding a little talk to my guardian angel with a request for help. I made my pledge on this night as well. If the cause of the accident was the result of the continued disregard for the known deficiencies of the performance of the ATR in ice, which I suspected, I would no longer be silent.

Tony Zarinnia was disturbed by what he swore was his best friend's voice on the radio calling for help as he maneuvered his Beech 1900 airliner for the eighth landing of the day back at the Central Wisconsin Airport. When he heard about the American Eagle ATR crash, he immediately thought of Jeff and tried to call the Gaglianos to make sure Jeff was alright, but the phone was busy. He got the bad news the next day and made a request of his boss at Skyway Airlines for the day off so he could go and be with the family of his friend. The response was from the typical regional airline manual: "You are scheduled to fly and if you do not show up, you are fired."

Later on Halloween night, Kim Collins, Sandi Modaff's sister, spontaneously miscarried after hearing about her sister's death. The morning of November 1, as investigators and police technicians entered the field near Roselawn, Indiana, to photograph and inspect the carnage, Sandi's sister was in a Pennsylvania hospital enduring another loss.

5

The investigative process

The National Transportation Safety Board is a small agency within the Department of Transportation of the federal government. Charged with the primary responsibility of investigating accidents and other safety problems in five modes of transportation, the NTSB employs about 350 people in roles of investigation, administration, and support. The Federal Aviation Act of 1958 created the agency, and the Independent Safety Board Act of 1974 expanded its legislative mandate, placing primary responsibility for investigating and determining the probable causes of all civil aviation accidents with the NTSB. Although it may delegate the actual investigation chores to other agencies, such as the FAA, the board is the only entity that can determine the official probable cause.

The 1993 NTSB annual report to Congress described the makeup and operation of the board. The board is composed of five members; the chairman currently is James Hall. Under the control of the chairman are four branches, or offices, staffed by the investigative, technical, and support staff. The Office of Aviation Safety fills a number of functions, including the primary responsibility for investigating aviation accidents and incidents, drafting reports, and proposing probable causes to the full board for a vote. In conjunction with the other board offices, the Office of Aviation Safety also formulates aviation safety recommendations addressing concerns outside of specific investigations.

The Office of Aviation Safety personnel are scattered throughout six regions and four field offices in major metropolitan areas across the United States. The six divisions that comprise the Office of Aviation Safety are Major Investigations, Field Operations and General Aviation, Operational Factors, Human Performance, Aviation Engineering, and Survival Factors. In the case of a major airline accident, like American Eagle Flight 4184, most of these divisions become involved.

As a result of the extensive media attention associated with an air carrier disaster, aviation accident and incident investigations operate in a public forum, and the credibility and independence that are required for the investigation necessitate a tightly controlled flow of information before the release of a final report. The board is free to develop its recommendations for safety improvements without resorting to a cost-benefit analysis, as do counterparts at the FAA, and it frequently irritates the FAA with its far-reaching and costly ideas. Normally the recommendations are included in the final report; however, in the interest of safety, the board may issue proposed improvements at any time during the investigation. Unfortunately, the NTSB has little regulatory power and can only provide these recommendations to the controlling agencies, which are free to accept, partially implement, modify, or disregard them entirely. Interagency rivalries between the NTSB and the FAA sometimes dictate whether the cause is accepted and proposed changes implemented. A toothless tiger, the NTSB is often frustrated by an inability to force the FAA to act, and often years or even a decade might pass before the FAA finally acts as a result of public pressure, often only after a disaster has occurred.

When the board is notified of a major aircraft crash, it launches a "go team," which varies in size depending on the severity of the accident and the complexity of the issues involved. The team consists of experts in as many as 14 different specialties, and each expert, in turn, manages other specialists from industry and government in collecting the facts, conditions, and circumstances surrounding the accident.

The participation of outside parties, such as the aircraft manufacturer, airline, pilot's labor union, or other authorities in various fields multiplies the board's resources and fosters a greater likelihood of agreement over the findings of the investigation; wide participation, however, also opens the door for the possible slanting of information supplied in the final reports. The cooperative efforts also allow firsthand access to developed information so that corrective actions may be taken by the appropriate parties in a timely fashion. At least that is the goal, and for the most part, the system works. Problems develop when the supposedly politically independent board allows the perception of behind-the-scenes maneuvering. To supplement the investigation, a hearing may be convened or depositions of knowledgeable parties taken under oath to collect additional facts. After the investigation is complete, the investigators prepare a detailed narrative report that analyzes the investigative record and identifies the probable cause of the accident and contributing factors. A major investigation can take from six months to more than one year to complete.

The investigator-in-charge is provided by the Major Investigations Division and manages the preparation of the board's aviation accident reports. The group chairpersons, under the direction of the investigator-in-charge, coordinate the efforts of industry and other government agency participants in the accident investigation. Each group is charged with conducting an accurate, objective, and thorough technical investigation of the accident and producing factual reports that are incorporated into the final report of the accident. The reports are likely to include proposed safety recommendations that the board is free to adopt or omit.

Operational factors experts in three disciplines (air traffic control, aircraft and airline operations, and weather) support major investigations with intensive work in their specialties. Air traffic control specialists examine air traffic control facilities, procedures, and flight handling, including ground-to-air voice transmissions, developing flight histories from Air Route Traffic Control Center and terminal facility radar records. Other specialists examine factors involved in the flight operations of the carrier and the airport and in the training and experience of the flight crew. Weather specialists examine meteorological and environmental conditions that might have caused or contributed to an accident.

Human performance specialists examine the background and performance of the individuals associated with the accident, including their knowledge, experience, training, physical abilities, decisions, actions, and work habit patterns. Company policies and procedures, management relationships, equipment design and ergonomics, and the work environment also are examined.

Aviation engineering experts in four areas provide technical investigative skills. Powerplant specialists examine the airworthiness of aircraft engines, and structural experts examine the integrity of aircraft and component construction, including flight controls and design and certification. Systems specialists examine the airworthiness of aircraft flight controls and electrical, hydraulic, and avionics systems. Maintenance specialists examine the service history and maintenance of aircraft systems, structures, and powerplants.

Survival factors experts investigate factors that affect the survival of persons involved in accidents, including the causes of injuries and fatalities. These investigators also examine cabin safety and emergency procedures, crash capabilities, equipment design, emergency responsiveness, and airport certification, if applicable.

The passenger seating capacity of the ATR-72 requires that American Eagle operate the aircraft under the same rules that American Airlines operates larger aircraft. The parameter that differentiates a commuter airline from the larger, better-known airlines is seating capacity of the aircraft, not the propellers. With fewer than 30 seats, airlines operate under Part 135 of the Federal Aviation Regulations; with 30 or more seats, operations are conducted under the more stringent Part 121. Public perception of the safety of commuter airlines was undermined by erroneous press reports listing American Eagle Flight 4184 as a commuter airliner. Although the statistics vary year by year, the safety rates for propeller airliners operated under Part 121 regulations are equal to the safety rates for jet aircraft under the same rules. Accidents are not a function of the type of powerplant that drives the aircraft or a result of the commitment to a "zero accident" philosophy. Nobody wants to have an accident, and every day of training, every hour of study, and every function during the day is designed to prevent an accident. Unfortunately, more accidents occur in some years.

Major regional airlines operating under Part 121 of the FAA regulations, which includes United, TWA, American, and Simmons Airlines/American Eagle, had only one accident, a fatality, in 1993. On April 4, 1993, on his first day back flying the line after a stint in management, first as assistant chief pilot and then as chief pilot at Chicago O'Hare, Simmons Airlines Captain Dave Peahl was at the controls of an ATR-42 preparing to depart Chicago for a night flight to South Bend, Indiana. The ground crew was preparing to close out the flight, and the crew had started the engine on the right side. A bag from a late-arriving passenger was delivered by an American Airlines baggage handler to the right rear cargo door of the ATR, and an Eagle ramp worker directed her to take the item to the left front cargo bin. Because a beacon light was inoperative on the bottom of the aircraft, the ramp worker saw no visible warning of the engine's operation, and she walked into the spinning propeller and was killed instantly.

For airlines regulated under Part 121, including American Eagle, 1994 was worse: 269 people died in six accidents, including 83 dead on two American Eagle commuter flights. Officially, the 15 fatalities from the crash of American Eagle Flight 3379 near Raleigh-Durham, North Carolina, in December were classified under a Part 135 airline accident because the Jetstream 32 aircraft had fewer than 30 seats; however, American Eagle operates all of its four separate airlines under the stricter Part 121 standards.

For record-keeping and reporting, AMR management maintains the four Eagle carriers as separate operations, because the group would be classified as a major airline if they were combined. A major airline is an operation that generates $1 billion in annual revenues, and those operations require that additional federal reports be filed.

A review of accidents with fatalities involving Part 121 U.S. carriers between 1976 and 1994 indicated no direct links between operations and accidents. In addition, as years pass and technology progresses, a decline in the accident rate has not occurred. During this 19-year period, only one year, 1980, passed with no fatal accidents involving air carriers. The previous year, 1979, was exceptionally bad, with four accidents claiming 351 lives; the following year, 1981, also had four fatal accidents with four deaths. The deadliest single year for the airlines was 1985, when the NTSB listed 526 fatalities in seven accidents; on the opposite end of the spectrum, aside from 1980 with zero deaths, 1993 had only one fatality in one accident. Coincidence and chance place any of the eight million annual Part 121 airline departures at risk.

Notification of the fifth airline disaster of 1994 came at 4:45 P.M. CST on October 31, 1994, to Gregory Feith, the next available investigator-in-charge. Feith and a go team departed Washington, D.C., later that evening on a Gulfstream IV corporate jet belonging to the FAA. The NTSB crew landed in Chicago and traveled by car to Merrillville, Indiana, 25 miles north of the crash site, arriving at 12:30 A.M. on November 1, 1994. NTSB Chairman James Hall was the board member on duty and was coincidentally in Chicago on other business. Hall joined the team, and an NTSB command center was established at the Radisson Hotel just off I-65 in Merrillville.

The team consisted of experts in a variety of specialties drawn from the headquarters and air safety investigators from the board's North Central Regional Office. Areas of expertise included aircraft operations, human performance, aircraft structures, aircraft systems, powerplants, maintenance records, air traffic control, survival factors, biohazard protection, aircraft performance, meteorology, and witness management. Laboratory services at the NTSB were notified to have specialists available in Washington, D.C., to conduct the readouts and analysis of the cockpit voice recorder and the digital flight data recorder as soon as they were found.

The response was immediate, and the early revelations were enlightening, but the in-depth investigation lasted for more than a year before an approved report was submitted to the board for a vote. The investigation was simplified by the state-of-the-art flight data recorder, which provided more than 100 parameter readings on everything from engine power to flight control positions. One complication was that the ATR-72-210 was manufactured in France. Reams of data were available to the investigative team, and each bit of data was intensely scrutinized and debated.

American Eagle Flight 4184 had been in a holding pattern at the LUCIT intersection for about 33 minutes before the crash. The airliner was at 10,000 feet for the initial hold and had been cleared to begin a descent to 8000 feet. After Flight 4184 was notified that the hold would last another 10 minutes, it went out of control. The aircraft had been flying a 10-mile, racetrack-shaped pattern, completing the circuit every 8 to 9 minutes. Aircraft N401AM passed over the little community of Roselawn, Indiana, four times before it dove into the farmer's field south of town. The official reason for the hold was reportedly reduced operations at Chicago O'Hare due to poor weather conditions and gusty northeast winds, which eliminated the use of additional runways.

The air traffic control group of the NTSB was responsible for investigating the air traffic control factors that might have played a role in the accident. Sandra Simpson of the NTSB chaired the team, which included representatives from the National Air Traffic Controllers Association, the FAA, American Airlines, and the Air Line Pilots Association (ALPA). The first duty of the group was gathering operational data from the various air traffic control facilities and controllers who had contact with the flight and procedural data to determine whether any actions violated established rules. Eagle Flight 4184 was operating in the parcel of Chicago Center airspace known as the Boone Sector. It consists of approximately 1400 square miles and extends from the ground to 10,000 feet. Aircraft en route to Chicago O'Hare from the southeast pass through the Boone Sector before they are handed off to Chicago O'Hare Approach Control for final routing to the airport. Under an agreement with Chicago O'Hare, the Boone Sector controller is to position the incoming aircraft at less than 10,000 feet in a descent for 8000 feet before passing control of the aircraft. Shortly before the flight was issued descent clearance, Chicago O'Hare told Boone it would take the ATR within 5 minutes, and in anticipation of the transition, Flight 4184 was cleared for the lower altitude.

In addition to its traffic separation duties, Chicago Center has a weather coordinator responsible for disseminating pilot report information to the various sectors. Numerous pilot reports were received by the center on October 31 pertaining to ice and turbulence, confirming the weather conditions and flight precautions. No pilot reports indicated the presence of any weather phenomena that were not in the forecast. At the time of the accident, a Saab 340 turboprop 47 miles to the east of LUCIT reported light-to-moderate rime ice from 13,000 through 16,000 feet and light chop through 19,000 feet.

The delay experienced by the flight on the ground in Indianapolis was the result of a national program administered at the Air Traffic Control System Command Center in Herndon, Virginia. The operation coordinates traffic flow across the country by eliminating or reducing bottlenecks caused by a high density of traffic or weather conditions. Flight plans for scheduled airline flights are stored in a computer database that is updated monthly. A team of specialists monitor isolated ATC restrictions and issue "flow control" flight delays, known as control departure times, as required. The presence of a control departure time requires that the aircraft be held on the ground when "delays are projected to occur in either the en route system or the terminal of intended landing" and ensures that a minimum of airborne holding will be required, reducing airspace congestion. The language of the rules is unclear as to whether the Indianapolis controllers should have kept Flight 4184 on the ground when told by flow control, "he is released. That fix is in the hold so he might have to do some holding when he gets up here, but he's released."

The air traffic control group traveled to Herndon to tour the Air Traffic Control System Command Center facility on December 20, 1994, and interviewed two individuals. NTSB had requested that the FAA save all documentation regarding programs in effect on the date of the accident, but the investigators found that the FAA had not passed the request to Air Traffic Control System Command Center until November 17, 1994, 2 days after the 15-day rotation period. As a result, all documentation had been destroyed.

The information obtained during the review of the documents might have helped the air traffic control group determine the appropriateness of the handling of American Eagle Flight 4184. Either because of the NTSB's delay in failing to request the paperwork until November 15, or the bureaucracy of the FAA, which forwarded the letter two days later, valuable documentation was lost. The investigators were forced to review pieced-together data and recollections of involved parties. Things were better when the group interviewed controllers at the Chicago Center in Aurora, Illinois, during the first days of November.

Air traffic control developmental specialist Michelle Willman was handling Eagle Flight 4184 when the accident occurred. Hired by the FAA on September 26, 1989, Willman was qualified at several air traffic control positions and was roughly three-fourths of the way through her training as a full performance controller at Chicago Center; although referred to as a trainee in press reports, she was not an inexperienced rookie. On October 31, she came on duty at 2:00 P.M., was scheduled to work an 8-hour shift, and had just come off of a 30-minute break before she and her monitor, John Hall, relieved Michael Borzym at the Boone Sector station. While training at a position, a controller is monitored by a qualified instructor who sits alongside the trainee, watching and instructing.

Michael Borzym had received Eagle Flight 4184 on the frequency and issued the Chicago altimeter setting before being relieved. No weather reports had been received during his shift at the Boone sector station, but he had information of light-to-moderate chop between 8000 and 10,000 feet and light icing at 10,000 feet. The radar screen showed a light green "9-inch-wide paintbrush pattern" that indicated light rain. When he was relieved by Michelle and John, Michael supplied a briefing on the traffic and computer problems that had been showing up that day. The computers had been "burping" all morning, meaning the radar scope would display a "not updating" message and the air traffic controllers would lose the entire radar display for less than a second. When the display came back on line after what was perceived as a flicker of the screen, it would take between 40 and 60 seconds before all data was back on the display. It was an annoyance, but not an unusual occurrence at the Chicago Center with its older computers. Michael told his relief that he had just been advised to "prepare himself for a possible hold" and that the only aircraft on frequency was Eagle Flight 4184 at 10,000 feet en route to the Chicago Heights VOR. No complaints from or extraneous conversations with the crew of the flight had occurred.

Michelle Willman took control of the flight, and soon after, issued the holding instructions. She attempted to keep the crew of the flight updated on the time expected in the hold but had some trouble getting accurate information from the approach controller, who advised "it was flows' restriction." By guessing, she issued the first expect further clearance time and then adjusted it because she felt it would not be adequate. The workload at Boone was light, but the complexity was building as the ATR circled over LUCIT. Another airliner, Kiwi Flight 017, was inbound for Midway Airport and conflicted with Eagle Flight 4184, so the decision was made to descend the ATR to 8000 feet to provide maneuvering room for the controller with the jetliner.

Shortly after she issued the clearance for descent, Michelle noticed the radar screen data block for Eagle Flight 4184 change to "XXX," meaning the computer could not keep up or could not determine the next altitude reply from the aircraft. Although she thought this was unusual, she did not do anything until she lost the radar track on the aircraft shortly thereafter. It was agreed she would issue the flight a "radar contact lost" report and attempt to determine the position of Eagle Flight 4184. While Michelle called the flight, John walked over to the supervisor position to report that they had lost radar and radio contact with the Eagle aircraft. Michelle was not overly concerned at the time and asked for assistance of other aircraft in calling and verifying the quality of her own radio transmissions. Finally, attempts were made by airliners in the area to listen for an emergency locator transmitter on 121.50 megahertz frequency.

John Hall reported that he thought the hold exceeding 30 minutes was "kind of long" and discussed a plan with Michelle if Eagle Flight 4184 could not continue to hold. Although he could not provide his definition of a time that equated to "too long," he stated that holding is usually shorter than 30 minutes. No other aircraft were holding in the Boone sector, and he was not told of the reason for the hold. Michelle and John were informed of the crash about 15 minutes after the airplane disappeared from the radar.

Michael Debb was the South Area supervisor at the time of the accident. He had been at the Chicago Center since 1970. John Hall called him to the sector station and advised him that they lost radar contact on Eagle Flight 4184, and the crew was not answering. After asking the controllers to call again, Debb obtained the telephone number for American Eagle operations and called them. Operations personnel told Debb that the flight was "in holding and they cannot talk to the flight." Debb told them to "try again right now." He then explained that air traffic control had lost radar and radio contact with the flight and informed Ken Legan, his own supervisor. The controllers showed him the last position on the radar screen, and the supervisors called the two closest airports, at Rensselaer and Lowell, Indiana, to determine whether the aircraft had landed at either location. The response was no at both airports. Michael Debb then called the Indiana State Police and was talking to Sergeant Lee when a report came in about a crash southeast of Roselawn. Debb called dispatch for American Eagle in Dallas and advised them the aircraft had crashed and obtained the number of people on board, 68. The controllers were relieved of duty and advised to write statements while the facts were fresh in their minds. Finally, Michael Debb took possession of the flight progress strip.

Nobody involved in the air traffic control area had any warning of anything out of the ordinary with either the weather conditions or the flight. By all reports, the forecast weather existed almost exactly as predicted, and no other aircraft reported problems. Other than the hold, the progress of Eagle Flight 4184 had not been unusual in any way. No warning signs were found that were ignored, and other than the release of the flight with an expected airborne hold, which is technically against FAA policy, no other violations of the rules were noted in the chairman's report of the air traffic control group.

On November 11, 1994, in response to an emergency recommendation by the NTSB, the FAA issued a general notice, or GENOT, that applied to all air traffic control facilities. It stated:

Air traffic managers shall determine which controllers in their facility are affected by this GENOT. For designated affected controllers this is a mandatory briefing item. This briefing shall be entered into all designated controllers' training and proficiency records, 3120-1, within 30 days of receipt.

On October 31, 1994, an ATR-72 was involved in a fatal accident. Pending a special certification review of the aircraft, ATR-42 and ATR-72 pilots have been advised to avoid extended exposure to icing conditions and fly at indicated airspeeds in excess of 175 knots while holding. The following procedures are effective immediately: ATC personnel shall provide priority handling to ATR-42 and ATR-72 pilots when they request route, altitude, or airspeed deviations to avoid icing conditions. ATC personnel should be aware that when the ATR-42 and ATR-72 use speeds in excess of 175 knots, they may not be able to remain within the confines of holding airspace.

Investigation later revealed that it was not convenient to work the turboprop into the inflow of jet traffic from the west sectors. While controllers watched for an opening, Flight 4184 entered the icing conditions first reported by the crew at 3:48 P.M. when a statement was made by the first officer, "I'm showing some ice now." Eight minutes and 47 seconds later, the aircraft departed from controlled flight and spiraled to the earth at 425 miles per hour.

The witness group of the NTSB was charged with locating and obtaining statements from witnesses to the accident. Stephen Wilson of the NTSB chaired the team that included representatives from the FAA, ALPA, and the Federal Bureau of Investigation (FBI). The group was formed at the time the on-scene investigation was initiated, and they discovered that the FBI and the Newton County sheriff's office had already begun a coordinated effort to identify known and potential witnesses and conduct interviews. An appraisal of the procedures already in use determined them to be "the most efficient use of resources and manpower necessary to fully develop the facts, conditions, and circumstances surrounding the accident."

During the on-scene portion of the investigation, 19 witnesses were located and interviewed. Six witnesses were classified as eyewitnesses, 13 as ear witnesses. As can be expected with the recollections of witnesses, a number of conflicting statements occurred, with some information later proven to be inaccurate.

Four of the six eyewitnesses were in motor vehicles at the time of the accident, two riding together and the other two in separate cars. One eyewitness was outside and another watched from a window in a residence. Two of the three motor vehicles were reportedly in motion at the time the operators caught a glimpse of the aircraft; one was stopped at an intersection. Of the six eyewitnesses, only four identified what they saw as an airplane; one thought it was a helicopter; the other said it was a small plane, but when he learned the size of the airliner, he was unsure if he saw the airplane or a part falling from it. Only one eyewitness properly identified the direction of travel of the aircraft as southwest to northeast. Four eyewitnesses described the plane's descent angle as 20, 30, 40, and 45 degrees nose-down. Two did not indicate that it was descending. None described the aircraft as rolling or inverted, although the possibility exists that two witnesses had the perception that the wing should be on the bottom of the fuselage; it is at the top on the ATR. Two witnesses stated the airplane was trailing "smoke," which was most likely water vapor.

The ear witnesses included seven who only heard a loud noise; four others said they thought it was thunder associated with the rain; one described engine sounds only; and one described a loud "flat" noise followed by whining engines that ended with a sound like a clap of thunder. Ear witnesses were located from a few hundred yards to three miles from the crash site. None of the witnesses had more than a couple of seconds to see or hear the aircraft, and their statements revealed little information beyond what would be determined during other phases of the investigation.

The area of the investigation that was subject to the greatest level of scrutiny and contention began on November 1, 1994, by the members of the meteorological group. This group was chaired by NTSB employee Gregory Salottolo and staffed by representatives from the National Weather Service, National Air Traffic Controllers Association, ALPA, French Bureau of Accident Investigations (also called the BEA), American Airlines, and the FAA. The group was given the task of defining the environment in which Flight 4184 was operating before and up to the accident and documenting the pertinent products, services, and actions of agencies and individuals involved in the accident.

Doppler weather data were collected from sites near Joliet, Illinois, and Indianapolis. Digital data from a geostationary operational environmental satellite were obtained and reviewed on the NTSB's Main Computer Interactive Data Access System workstation. Information was provided by the National Weather Service, FAA, and American Airlines, which operates its own in-house weather forecasting and monitoring facility that produces data rivaling the quality issued by government agencies. The amount of data collected was impressive and presumed accurate, but the correlations to what the crew observed on the aircraft could have different interpretations.

The 4:00 P.M. National Weather Service weather depiction chart showed a low-pressure center in west central Indiana. Cloud ceilings of less than 1000 feet or visibilities of less than three miles in rain were occurring throughout the area of operation for Flight 4184 in northern Indiana and southern Illinois. A moderate cold front extended from the low-pressure center to the southwest, and a moderate stationary front extended eastward from the low-pressure center. Winds around a low-pressure center flow counterclockwise. The accident location was nearly at the center of the low, which meant the ATR was subjected to changing wind directions as air flowed into the area from the northeast bringing moisture off Lake Michigan, from the northwest bringing colder air, and from the south bringing warmer air. The confluence resulted in high-moisture air with the potential for significant lifting and cooling. It was described later as a strong storm, but by no means the storm of the century. By all accounts, weather forecasts had been fairly accurate, but it is not uncommon for unique icing situations to be concentrated and itinerant.

At the time of the accident, observations at the Indianapolis Airport, 93 miles to the south of the crash site, revealed the weather was mild with a temperature of 64 degrees with a 20,000-foot overcast cloud deck and a broken layer at 9500 feet; the winds were out of the southeast at 9 knots. However, Lafayette, Indiana, 49 miles northwest of Indianapolis and 44 miles southeast of the impact, had a 1000-foot overcast ceiling and a broken layer of clouds at 600 feet. The temperature was 46 degrees, and winds were out of the northeast at 16 knots. At Gary, Indiana, 32 miles north of Roselawn, conditions included a temperature of 44 degrees with a scattered layer of clouds blown in from the lake at 800 feet and a ceiling of 1700 feet; winds were from the north-northeast at 13 knots with gusts to 30 knots. Although the conditions on the ground revealed a moderate weather day, they also indicated a changeable and worsening condition centering in the northwest corner of Indiana, exactly where Flight 4184 circled.

Upper air winds measured by the Doppler radar sites showed a wind shift of 90 degrees between 8000 feet, where winds were out of the east at 25 knots, and 11,000 feet, where winds were out of the south at 25 knots. Temperature profiles were retrieved from a jetliner flying from Chicago O'Hare to Boston about 50 minutes after the accident. The recorded data showed a consistent, although shallower-than-standard gradient, decrease in temperatures with altitude and placed the temperature at 10,000 feet at minus 6.2 degrees Celsius, or 21 degrees Fahrenheit. The same aircraft had the capability of recording upper-air winds encountered during the flight and indicated similar direction and speeds as interpreted from ground-based weather radar data. Information supplied by the National Lightning Detection Network indicated no cloud-to-ground lightning strikes in the vicinity of the accident during a 10-minute period on either side of the crash time. Rainfall from ground stations reported intervals of no recorded rainfall and periods of light rain during automated measuring over spans of 15 minutes.

The collected satellite data indicated an area of higher cloud tops that moved from west of the crash site to the northeast during the time the aircraft was flying in the holding pattern. Between 3:32 P.M. and 4:02 P.M., the estimated cloud heights varied from 29,200 to 16,900 feet. Cloud movement was to the northeast.

The weather data meant nothing to the crew of Flight 4184 or other aircraft flying through the area of weather and is best represented by pilot reports of being "in the soup." A small, twin-engine business aircraft flying at 12,000 feet reported light rime ice over the Boiler VOR just northwest of Lafayette, Indiana, at 3:10 P.M., and a Saab 340 turboprop airliner indicated the icing over the same location as light-to-moderate at 15,000 feet about seven minutes later. Eagle Flight 4184 passed through the area between 14,000 and 16,000 feet about the same time but reported no ice to controllers, and no discussion regarding ice is known to have occurred between the crewmembers. Eighteen pilot reports were collected within 100 miles and one hour on either side of the crash, and none reported anything worse than light-to-moderate rime or mixed icing.

Meteorologists on the ground stated that between 2:00 and 4:00 P.M. there "were no pilot reports of any significant weather in the area of the accident." According to the meteorologists, a shear layer was evident in the area, indicating a horizontal level where winds changed direction or speed dramatically, but the layer was a steady-state feature and changed little throughout the day. Aircraft had been flying through the layer all day, and its existence did not trouble the weather prognosticators. The layer was located between 8000 and 11,000 feet and only caused concern because of the direction change and not the wind speed, which remained consistent.

Simmons dispatch monitored the weather situation, the dispatcher stated, and the area weather was not a cause for concern because the forecasts had been accurate and pilot reports showed no disturbing trends. No call was received from Eagle Flight 4184 to verify weather conditions, possibly indicating that nothing untoward was visible from the cockpit. The weather group began to focus on the possibility that a freezing rain or freezing drizzle might have caused ice to accumulate on the aircraft, a hazard of which the crew was not aware.

The survival factors group of the NTSB had little to investigate because the destruction of the aircraft was so complete that no survivors were possible. Debbie Childress of the NTSB chaired the team, which included representatives from the pilot's and flight attendant's unions, the FAA, American Airlines, the FBI, the Indiana State Police, and the Newton County HazMat Coordinator. The report was brief: of the 64 passengers, 46 were adult males, 21 were adult females, and one was a male child, age four.

The structures group of the NTSB was responsible for overall documentation of the accident scene and aircraft structure. Frank Hilldrup of the NTSB chaired a team that included representatives from ALPA, Simmons Airlines, the FAA, and ATR Support, Inc., a U.S. corporation established to provide a liaison between the aircraft manufacturer and North American operators.

The destruction of the aircraft was complete; small pieces were scattered throughout the debris field. Two large pieces of the fuselage were easily spotted on the eastern edge of the spray of aircraft parts and human remains; the tail cone section, with the vertical stabilizer still attached, lay in the mud. The cockpit voice recorder and flight data recorder, often referred to as black boxes, were stored in the tail cone, and from the wreckage, it is obvious why. The tail cone, mostly intact, was the largest piece of debris. A heavy landing gear assembly had been catapulted to the far northeast corner of the debris field. The summary of the structure group's report tells the story.

> *Although damage to the airplane precluded a complete accounting of the airplane structure, portions from the forward and aft ends of the fuselage and both wingtips were located at the site. The forward fuselage and cockpit had been destroyed. The Structures Group reassembled recovered portions of the wings, horizontal stabilizer, and flight controls to assist in accounting for these areas of the plane. Portions of all flight controls were found in the debris field; some were located nearly 1000 feet from the impact craters. Examination of the flight control stops revealed damage consistent with overtravel of the control surfaces. No ice was observed on the wings or empennage on arrival at the scene the morning after the accident.*

Note that ice would not be expected to last 17 hours in above-freezing temperatures. Three craters were located in the field, and most of the aircraft debris, except for those portions that separated before the crash, were located in a 60-degree arc to the east and northeast of the impact points. An outboard section of the left wing was found south of the crater. The horizontal stabilizer was found west-southwest of the impact craters, and portions of the left elevator and left aileron were found southwest of the creek bordering the field. The counterweights for the elevators, ailerons, and the rudder, plus portions of all flight controls and doors were found in the debris field. After the airplane wreckage was removed from the edge of craters, water was drained from the depressions, and visible pieces were again removed. A backhoe was then used to excavate each crater. Excavation removed about three feet of soil below the bottom of the craters, and the soil was examined to identify and remove airplane parts.

The structures group confirmed that the ATR was nearly inverted at impact; right engine components were found northeast of the craters and left engine components were found to the southwest of the main impact point. Fuselage, insulation, wing structure, and human remains were found in the largest crater. Components and wiring from the cockpit were found near the center of the hole.

Before impact, the outer 10 feet had separated from both wings, but because the wings are constructed of a composite material, it was impossible to determine the directions of the failure. The outboard piece of the left wing was found 390 feet to the south-southwest of the southwest impact crater. The outboard section of the right wing was found 80 feet east of the eastern edge of the large crater. All of the wing-to-fuselage attach fittings for the center wing section were found, although the section was broken into numerous pieces. An upper panel from the wing center section was found 800 feet southwest of the impact points.

The horizontal stabilizer was found intact 165 feet west of the southwest crater. No visible damage to the leading edge or the deicing boot was observed other than a small puncture to the lower surface of the left half of the stabilizer. A deformation of the aluminum skin of the component was found, and the left lugs used for attachment had pulled through the bottom of the attachment points, indicating tensile overload.

Most portions of the recovered fuselage were found east-north-east of the impact craters, although some pieces of forward fuselage and insulation were found in the largest impact crater. The least damaged areas were from the aft area of the fuselage. A six-foot section of the aft fuselage was found 150 feet east-southeast of the tail cone section and included the belly skin and left skin to the crown of the cabin. A right-forward piece of fuselage skin with the lettering "n E" of American Eagle was found in the large crater and had been compressed in an "accordion" fashion in a mostly fore-aft direction.

All doors were determined to be closed; "witness marks" on the latches and the pins that secure the doors were found in the proper extended position. One of the main landing gear actuators was found in the retracted position, which is consistent with landing gear in the up-and-locked position. Although little was left for a metallurgical study, all examined surfaces contained no prior-to-accident fractures.

The powerplant group's investigation consisted of three phases. Phase 1 was performed November 2 to 4 at the Roselawn crash site, phase 2 consisted of a propeller teardown investigation at the Hamilton Standard plant November 17 and 18, and phase 3 involved the engine teardown examination at the Pratt & Whitney-Canada facility December 6 and 7. Edgar Fraser of the NTSB chaired the team, which included representatives from ALPA, Simmons Airlines, FAA, Pratt & Whitney, Hamilton Standard, and several Canadian aviation officials. All components of the powerplant were severely damaged but identified at the crash site. The composite propeller blades were shattered and scattered around the site, but pieces were recovered from each of the eight blades, indicating no inflight loss. Rotating parts in the engine showed circumferential rubs, indicating that the engines were operating at impact. There was minimal evidence of postimpact fire on the engine parts.

All recovered parts were removed to a cleaning area where biohazard decontamination was completed. Parts were then packed and shipped to the facilities of various manufacturers. Even though damage was extensive, no evidence pointed to engine or propeller failure before impact.

The systems group of the NTSB was responsible for inspection and investigation of all aircraft systems for operational details and confirmation of the proper inflight utilization and performance. Robert Swaim of the NTSB chaired the team, which included representatives from ALPA, Simmons Airlines, FAA, French BEA, and ATR.

This group faced a mountain of work that included the verification of all onboard systems design, problem history, postcrash component inspection, and certification requirements. The team performed detailed testing on the roll, pitch, and yaw control systems; wing flaps; autopilot; ice-protection design and operation; computer components; instrumentation; fire protection; hydraulics; fuel; electrical; oxygen; pressurization; and landing-gear systems. Information developed in other areas of the investigation led the team to consider two important components of the aircraft as possible factors in the crash: flap control and deicing systems.

Recorded flight data indicated the upset began within a second of flap retraction. In a control emergency, pilots reverse the last action completed before the emergency, making flap control suspicious. Although the flap control and indicator could not be found, the group did investigate the flap control switch unit, which, although extensively damaged, revealed an internal configuration consistent with an attempted placement of the flaps at the 15-degree position. Unfortunately, the aircraft's computer prevented the flaps from being lowered at speeds above 185 knots, which was quickly exceeded. However, even with the airborne emergency, the flight crew had apparently attempted the reconfiguration.

The human factors group was chaired by Malcolm Brenner of the NTSB; he was assisted by staff from Simmons Airlines, ALPA, and the Civil Aeromedical Institute at the FAA. The group focused on issues relating to the pilots by conducting interviews of individuals familiar with Orlando Aguiar and Jeff Gagliano, and the furnished report failed to unearth any problems in the lives of either pilot.

The maintenance records group impounded the maintenance records pertaining to N401AM at Simmons Airlines. Group chairman W. Mitch Robbins led a team from ALPA, Simmons Airlines, and the FAA. At the time of the accident, the maintenance department had entered all data into a computer tracking program that was current for the aircraft through October 22, 1994. Aircraft operating times were accurate to October 30.

The original aircraft maintenance log for N401AM was established in Toulouse, France, February 2, 1994. The log showed installation of new engines on November 23, 1993, and new propellers on February 2, 1994. Factory testing consisted of 4.3 flight hours and three landings. On March 23, 1994, an aircraft inspection report acceptance sheet was signed by the Aerospatiale quality manager. The registration certificate was issued to AMR Leasing Corporation on the same day. On March 28, 1994, the airplane was placed on the Simmons Airlines continuous airworthiness and maintenance program; it entered service March 29, 1994. A review of aircraft records revealed no evidence of deficiency in compliance with airworthiness directives, service bulletins, or engineering orders.

A number of aircraft documents were recovered at the accident site, including an aircraft maintenance log for October 20 through October 31, 1994, a daily aircraft crew log, an ATR-72-212 aircraft operating handbook open to "Takeoff 47,000 Pounds" on page 27, a quick-reaction handbook that was closed, a Chicago sectional aeronautical chart that was unfolded and fragmented, and a maintenance item control sheet.

Two deferred maintenance items were in place at the time of the crash—"no. 2 bleed valve inoperative" and "door warning system inoperative"—and the configuration deviation list contained an entry indicating "right aileron wick(s) missing." All other maintenance items were in order.

Investigation of the two black boxes found at the site began with transportation of the sealed units to the NTSB laboratory in Washington, D.C., on November 1, 1994, where the protective covers were removed and the internal tapes recovered for transcription.

The cockpit voice recorder is one of the two black boxes often referred to in press reports; the other is the digital flight data recorder. In the ATR and most commercial aircraft, these units are carried in the tail cone section because it is considered the most survivable location on the airplane. Photographs of the crash site taken the next morning showed this tail cone section almost entirely intact with a large section of the vertical tail containing the American Eagle logo lying a distance from the main impact crater. The cases of the black boxes are painted orange with reflective striping for ease of recognition and location, and each is about the size of a large fisherman's tackle box.

The cockpit voice recorder records verbal communications, aural indications, and background sounds that can aid investigators in determining what exactly might have happened in an accident or incident. The unit records on audiotape that runs in a 30-minute continuous loop. Conversations of the flight crew, air traffic control, and other air-to-ground or air-to-air transmissions, public address announcements, and even the idle chatter between pilots and flight attendants are all captured on separate channels of the tape. The digital flight data recorder records the performance of the aircraft in snapshots of one-second intervals. This unit transcribes the last 25 hours of flight time on continuous-loop magnetic tape, with new data erasing the oldest data. The ATR's digital flight data recorder monitors more than 100 performance parameters and provides greater information than those data recorders found on some older jet transports.

These data and voice tapes contained chilling implications for flight safety due to the rapid onset of the violent roll control anomaly. Aircraft performance studies were intensive and became the central focus of the investigation.

Within days of the accident, NTSB Chairman Jim Hall issued a statement that a credible cause of the upset had been theorized; that theory suggested structural icing as a primary cause of the crash.

6

Recovery and identification

When the call went out for help after the report of an airplane crash near the small town of Roselawn in Newton County, Indiana, people responded in such overwhelming numbers that some had to be turned away. As the surrounding area swelled with out-of-towners requiring food, lodging, and physical and mental care, local residents opened their homes and kitchens and volunteered their services. The tiny collection of homes and farms quickly became known around the world. Before the investigation was over, nearly 1000 people had played a role in the recovery and identification of the victims. They asked for nothing in return for their acts of humanity, but the residents of Roselawn earned the respect and gratitude of the families and friends of the 68 people lost to tragedy.

Kathy Wann was spray-painting her husband's hair fire-engine red when the telephone rang just after 4:00 P.M. on Halloween. Carol Prohosky was on the other end of the line. Carol reported that her husband, Norman, had yelled to her to call the Lincoln County Volunteer Fire Department as he and the hired hand disappeared into the fields to find the downed aircraft. One of the first thoughts through Kathy's mind was that it was probably the local farmer who flies an ultralight aircraft and many times winds up walking home. This scenario had become almost common, and Kathy's husband, Cap Wann, chief of the the volunteer fire department, would give the man a ride home. He was usually irritated and sometimes scuffed up but otherwise unhurt. "But on an afternoon like this?," Kathy remembered thinking. Rain had fallen most of the day, even though it had slowed a bit in the past half hour; the temperature was barely 40 degrees, and the winds were gusting out of the north at nearly 40 miles per hour. Both Kathy and Cap did not believe any airplanes would be flying today, let alone the local farmer and his ultralight aircraft.

Cap Wann's hair was flaming red, and he was wearing a sweat-suit modified into a devil's costume, complete with homemade tail and cape. Cap responded as he had for decades. He headed for the station to pick up a rescue vehicle and raced toward the area of the reported crash.

Norman Prohosky's farm is less than five minutes down Highway 55 from the volunteer fire station in Thayer. Cap Wann was in a command vehicle he had picked up from the village garage, and he was joined by Bob Mauck, Dave Muilenburg, and John Knapp, the fire-fighter who had previously requested air disaster training from the FAA, in a second truck. The call had come in at 4:01 P.M., and by 10 minutes after the hour, the rescue team was in the area of the accident, straining to see through the mist and approaching darkness for wreckage. The station dispatcher had also been busy, contacting the sheriff's department and requesting the dispatch and standby of nearby medical units.

American Eagle Flight 4184 crashed near the southwest corner of a 40-acre plot that was not bordered by a road. The recently harvested soybean field had no direct access; it was bordered on the west and north by deep drainage ditches filled by the recent rains. The aircraft had cleared a stand of trees on the southern border of the field, leaving large sections of its wings and horizontal tail in adjacent fields along the flight path before plowing into the soft earth.

Robert Stone was deer hunting when he heard the breaking of the aircraft and subsequent accident. He described the sounds as a loud "plop" followed by the whining scream of engines and a final clap of thunder. The stresses on the airframe broke sections of the wings off the airplane. The ATR-72 wing has a higher rigidity than the swept wing of a jetliner, but rapid failure can occur when it is subjected to extreme loads.

Little of the airliner was left, and at the time the rescue crew did not realize they were looking for a large aircraft. Near the intersection of Highway 55 and County Road 700, which ran east-west along the north side of the field adjacent to the accident site, Chief Wann encountered a farmer who directed the group, which had tried to enter the field from the north but were stopped by the drainage ditch between the two separate tracts.

Clarence Hanley heard the emergency call over the scanner he kept at home. From his kitchen window, he could see the field where Flight 4184 crashed and the road from which Chief Wann and the other firemen were trying to reach it.

A low-flying plane sped over the home of Robert and Lynn Stone as Lynn and her children were preparing to go trick-or-treating. Lynn heard a loud crash and ran to the door and said, "That sounded like a plane crash." Her children passed her and ran around the yard looking for the source of the noise. Not able to spot anything, the family climbed into their minivan and started toward town. They spotted the emergency vehicles trying to get to the field off County Road 700. "That plane crashed, and I bet you any money that's where they're going," she told her kids. Driving her van around the firemen to stay out of the way, Lynn found the entry to the field east of the crash site. Seeing that the rescuers would be cut off by the drainage ditch, she drove in until the van got stuck in the mud and waved the firemen around to the entrance. "Go around this way," she hollered just as Clarence Hanley reached the group out by the road to take them around by the Stone family.

Bonnie Clinton is an emergency medical technician with the Shelbyville Fire Department. She was home with Rick, her husband, when they heard the call about a plane down at nearby Roselawn. Bonnie and Rick knew time could make the difference to someone who was seriously injured; if a plane was down, injuries were certain. Their trip to the site was quick, and the pair arrived to follow the firefighters down the muddy path near Lynn Stone, who was still signaling.

Several firefighters ran into the muddy field followed by the Clintons. Chief Wann saw the tail section of the ATR-72 lying about 400 yards to the south and thought a small airplane had rolled itself into a ball. As he closed on the wreckage, he saw a large tire from the aircraft to his right and realized that he had a major problem. The heavy rains had created a mire, and the rescue team sank and slipped in the ankle-deep muck. Bob Mauck had reached the main debris field first and came across the body of a small child and the scattered remains of others. It was quickly evident that there were no survivors.

Bonnie and Rick Clinton entered the area shortly after Bob Mauck and began searching for any sign of life. When she saw the red, white, and blue paint on the tail section lying a distance away, Bonnie thought, "My God, this is a commercial plane. But where are the seats?" The impact created two craters at the southwest edge of the debris, and Bonnie and Rick made their way toward them, thinking the fuselage of the aircraft might have buried itself in the depression. Everywhere they saw parts of human bodies, debris, clothes, and papers in the mud. That many bodies had to have come from a large plane, but where was it?

Chief Wann went back to his vehicle to call the Newton County Sheriff's Department and inform them of what was needed. It was still daylight, but darkness was approaching quickly. A large aircraft was down. It was 16 minutes after the initial call had come in. American Eagle Flight 4184 had been down for 20 minutes.

A search was conducted in the waning light. Once the sun went down, it would get dark under the low clouds, and Bob Mauck wanted to be absolutely certain they did not miss something. Mauck is the Lincoln County hazardous materials coordinator, and he declared the site a biohazard on the basis of the large amounts of biological and other possibly hazardous materials. It was the second time that an airline accident had been classified as a biohazard; the first was the USAir Flight 427 accident near Pittsburgh one month earlier. Bob quickly realized that the people walking through the carnage were not properly protected and facing danger; he also knew his turnout gear would have to be destroyed. In addition, the emergency was over, and the stress of wanting to do something was showing on the faces of those around him. By the time the search crew reached the southwest corner of the field, they had encountered 15 partial torsos and hundreds of body parts. Bob Mauck radioed Chief Wann to tell the sheriff's department there were no survivors. Bob remained in the field for another half hour, and at 4:45 P.M., the search was called off, and the sheriff's department and Indiana State Police secured the area for the night.

One hour later, Indiana Governor Evan Bayh ordered the search teams back into the field for another sweep. The sights were horrible in the daylight, but when illuminated by the beam of a flashlight, the scene was nightmarish. Other agencies and volunteer organizations began mobilizing, and a command post was established at the local fire department. The building had been set up for the Halloween bingo party and a craft bazaar. When the fire personnel arrived back at the station around 6:00 P.M., the Lake County Chapter of the American Red Cross was already on site and beginning to prepare to serve meals and providing counseling and stress relief.

The county highway department was called and asked to build an access road to the field; they worked through the night with portable lighting. Dr. David Dennis, the coroner, was called at 4:23 P.M., and before 5:00 P.M. the Indiana Funeral Director's Association was notified of a major mobilization. The FBI provided several agents for witness interviews and collaboration with the identification process. The local high school gymnasium was used as a temporary morgue; later the morgue was shifted to the National Guard Armory in Remington. The change salved the sensibilities of the locals, but it also meant the remains had to be transported over county lines, which could result in problems of jurisdiction. Dr. Dennis was assured by Indiana Governor Evan Bayh that he would have the governor's full support in anything he needed to do to expedite matters. In Indiana, the coroner is responsible for the accident site and morgue, identification of the bodies for the purposes of release to the families, providing a legal certificate of death, and proper disposition of unidentified remains. Dr. Dennis had authority over the recovery and final arrangements for remains not identified and sent to the families. Dr. Dennis is a dentist elected to the office of coroner, but he had never confronted what waited in the soybean field. He had the resources and cooperation of local, state, and federal authorities and the assistance of hard-working and dedicated volunteers. However, those who had walked among the scattered human and aircraft remains that afternoon did not comprehend the chore that awaited them the next day.

Most people assume that the first thing to be done after an aircraft crash is to remove the human remains. That is not the case. Questions about the nature of the accident and the enormity of the destruction existed, and sabotage or a terrorist act could not be ruled out immediately, so the decision was made to treat the site as a crime scene. The location and condition of nearly every human remain was important; location would be a monumental task because the body parts of 68 people were scattered in an area of about 8 acres. During the evening, the first of the aircraft's two black boxes had been recovered; the second was found the next morning.

Early on November 1, a meeting was held in Merrillville, Indiana, 20 miles north of the crash site, to coordinate and schedule the operations that must overlap without causing the loss or damage of critical evidence, either criminal or causal. Some of the NTSB team members had been at the USAir crash site the previous month and were prepared for the destruction in the soybean field. At the Pittsburgh crash, recovery workers picked up approximately 1600 body parts from the 132 passengers aboard the Boeing 737 jetliner. At Roselawn, for the 68 people on board Flight 4184, the number of body parts recovered exceeded 2000. Flight 4184 was the most destructive airline accident ever witnessed by the team from NTSB.

The FBI and Indiana State Police laboratory technicians were trained professionals, but the group also included volunteers who had never before participated in anything of this magnitude. During the meeting, the NTSB introduced the Incident Command System, which separated the tasks into manageable categories, with Dr. Dennis in a shared command role supervising the overall project along with the NTSB, the hazardous materials coordinator, and the Indiana State Police. A similar plan was developed and implemented at the morgue.

The Incident Command System made a difficult job manageable, if only slightly less frightening. Logistically, the team had to determine how to handle the biohazard issues; how to store the body parts before and after identification; how to transport the remains to Remington; how to communicate in the rural area with limited telephone capabilities; where to locate the command post; how to provide security at the site, en route, and at the morgue; how to screen the volunteers who would work in the field; what procedures to use for the identification process; how to cope with the inevitable visits by family members and curiosity seekers; and even how to prevent scavengers from desecrating or carrying off the remains. Supplies that needed to be procured included rolls of plastic sheeting; two-by-four lumber for sawhorses to make tables in the morgue; brown paper to cover the tables and line the walls and floor; plastic bags for the remains of passengers; gowns, masks, goggles, and gloves; and caskets for the final release to the families.

Communication trailers were brought in from Chicago O'Hare and the Indiana State Police. The decision to use the high school as a morgue had been changed; the building, however, was the only place large enough to serve as a command center. The two pay phones at the school and local phone service were stretched to the limit. Cellular phones became almost useless as a result of the volume of callers vying for a limited amount of cell capacity in the area. The authorities became concerned over the ability of local residents to contact fire, police, or medical personnel. Adding to the congestion, relatives called from all parts of the globe seeking information on their loved ones. Finally, many calls needed to be secure, and the agencies that came with their own satellite communications gear were glad they did.

Volunteers jumped into action the night of the crash, procuring and serving hot food and coffee to the firefighters, county workers, and police officers. One group of local emergency medical technicians found themselves unable to help with the site work because they didn't have the proper immunization for hepatitis, a form of which can be transmitted through infected blood.

Three volunteers who couldn't work at either the site or the morgue began helping by doing anything that was needed. Helen Mudd and her children, Julie Gunter and Chuck Mudd, canvassed three counties for groceries, always returning with more than the cooks had asked for. If the floors needed mopping or dishes needed washing, they pitched in. When the food and coffee was prepared, they hauled it out to the crash site. Their efforts were not unusual or unique, but they continued to help long after the disaster was over, forming a group called Mission Un-Impossible, which became a great comfort to grieving family members.

Federal authorities needed to search for critical components of the aircraft before the area could be opened to the recovery teams. Helicopters were already in the air recording the devastation with video and still cameras. The exact location of the impact sites were charted with sensitive global positioning devices, and measurements were taken between the large components, of which there were few. The FAA closed airspace around the crash site to prevent the media and curiosity seekers from becoming a hazard to those on the ground. State police equipped with binoculars recorded the N numbers of aircraft that violated the restriction.

The perimeter was secured, operational plans were put into place, voluntary response was overwhelming, communications were established, and laboratory and mortuary staff were called in from around the state and around the country. A physical anthropologist from Hawaii who had worked on the identification of missing-in-action service personnel from Vietnam was in the area on vacation and secured an extension from his supervisor to remain until this job was complete. With the severe level of "anatomical separations," identifying one body part from another was an important task. Identification of the dead was achieved by fingerprints or dental records. The FBI set up a computer system to record victim information provided by the families and log recovered items or identifiable remains; the walls of the morgue room contained a more primitive but still useful alternative.

Brown wrapping paper was taped to the wall to protect surfaces and ease the cleanup when the work was complete. On the floors, sheet plastic was laminated to the paper, creating a dangerous walking surface but adding to the ease of cleanup. On the walls, two 20-foot sections of the brown paper were divided into 68 1-foot square blocks to create ante mortem and postmortem lists that became known as the "tote boards." Information from the family such as clothing, jewelry, possessions, and physical descriptions were listed on the ante mortem list and matched with items found at the crash site on the postmortem list. As the amount of information on the brown paper grew, the workers became familiar with the individuals on the flight. The familiarity did not mean the treatment of the remains was any more or less professional or respectful; it just served to form an emotional attachment between the morgue workers and the victims.

The jumbled mass of agencies, organizations, and individuals were slowly forming into a unified team, mobilizing for a short-term project with long-term responsibilities. The work had to begin promptly, proceed expeditiously, and provide final results that were conclusive and accurate. Preparations progressed for receipt of the first remains scheduled for delivery to the morgue on November 2.

At the site, NTSB representatives collected the second black box on November 1 and shipped both recorders to Washington. Later in the day, police laboratory technicians began to establish a grid and place stakes by the remains. The organization of the site would make data available to identify as many people as possible. Bundles of wooden lath, orange ribbon, and small red body bags were delivered to the crash site. Normal body bags were not required. Investigators decided the lack of aircraft debris meant that most of the aircraft and passengers might have been buried in the main impact crater, and an industrial vacuum truck was contracted to pump out the water, body fluids, and jet fuel that had collected to determine exactly what was in the depression. Two helicopter loads of yellow protective coveralls, gloves, and boots were delivered to Roselawn. Indiana State Police laboratory technicians, trained to handle crime scene evidence and experienced in working around bodies in diverse conditions, were each assigned a group of five or six workers.

Workers received training in the procedures necessary to avoid transmission of disease. News reports of the training inflamed both the public and family members by implying that their relatives might have carried communicable diseases, to the dismay of Bob Mauck, the site's hazmat coordinator. Although some concern was placed on blood-borne pathogens such as human immunodeficiency virus, or HIV, and the potential of transmission, the main concerns were exposure attributable to organisms that develop on dead human tissue. No one approached any of the families to determine whether a passenger had any communicable diseases. Because of the impossibility of identifying all of the remains, it mattered little.

David Allison of PWI Environmental of Indianapolis jumped in to help with the shortage of protective gear. His firm was contracted to provide the pumper trucks and other equipment for the removal of water and waste from the impact craters, and he quickly assembled decontamination trailers, protective clothing, and other support. PWI crews ensured that investigative teams entered and exited the site at a single point so that decontamination procedures would be effective. Openings in overalls or seams between the gloves or boots and coveralls were taped to comply with established biohazard protection standards. Workers could not exit the area and return without going through the decontamination facility, so most endured the long days in the field without rest. Despite the training and precautions, 30 exposures of the workers to potential biohazards were reported.

Rick Schlegel of the Northeastern Indiana/Allen County Medical response team described the scenario in the after-action report prepared by the Indiana State Emergency Management Agency:

You need to remember that we were in the biohazard environment, and biohazard is easily transferred by the flies and the things we pick up. Bacterium grows independent of anything else when human tissue is exposed. I am concerned that some of you may think that biohazards are limited to someone who could have a communicable disease and was possibly on the plane. But that was not the concern. Later on, it did become a concern; however, if you assume appropriate precautions, you are covered anyway. But every single piece of tissue or any other type of specimen that is open to air will immediately start to grow bacteria. That bacteria is easily transferred. Now, right outside the door, there were flies around the trash cans that held disposables from your meals, and right there was the food wagon. If you developed any kind of a bacterial parasite, and you got ill, it would not be suitable for the medical team to also be ill. So, for that precaution, we always bring prepackaged food and we try very hard to stay within the disciplines of the MRE [meals ready to eat].

With both the crash site and morgue set up, the job began. On November 2, the weather moderated and the sun brought slightly warmer temperatures. The smell of 500 gallons of jet fuel was replaced with a pungent but highly recognizable odor. Once encountered, the smell of death is never forgotten. The description of the odor was a recurring theme in the remembrances of those on site. Months later, after the field had thawed from the winter freeze, local residents complained of the fumes still emanating from the field.

The process followed the same simple methodology more than 2000 times. A team member located a body part and alerted an official photographer by placing a 3-foot long, numbered stake into the ground next to the remain. The photographer then took several pictures of the remains. After the documentation was complete, the remains were placed in a red plastic bag with the stake number written on it; the bag was left at the wooden marker. Collection crews removed the bags and took them to refrigerated trailers. Coates Trucking had donated the use of several new refrigerated trailers for transporting and storing the remains until they could be processed, knowing the expensive rigs would have to be destroyed after their use.

When the morgue was ready, the trailers were hauled 40 miles to the Remington National Guard Armory under the watch of the Indiana State Police. At the morgue, the bags were opened, examined, and photographed again, and an attempt at identification was made.

Indiana State Police laboratory technician Rick Grisel, from Lowell, less than 10 miles from the crash site, led a team of volunteers.

Because the area was to be processed as a crime scene, the groups proceeded with caution. Soon, the concentration of stakes that marked body parts began to interfere with the investigators' ability to move through the crash site. The decision was made to mark groups of body parts with a wooden lath and associate all body parts found in the area with the number that was assigned to the marker. That activity became confusing to some family members, who later viewed the photographs and assumed the stakes represented their family members. Few body parts from the same person were grouped together. At the morgue, the workers could not assume that a foot and a hand found next to each other both belonged to the same person, even though both were labeled and photographed with the same stake number.

The first priority for the search groups was to pick up everything that might be a body part. The second priority was to recover certain parts of the airplane, such as cockpit instruments, that might become visible as the crews searched. The final priority was to pick up personal effects and remaining debris. That ranking left certain personal items in the path of the groups entering and departing the debris field for weeks.

The identification process was planned to be meticulous and designed to err on the side of no positive identification. Dr. Dennis anticipated the positive identification of no more than half of the 68 passengers and crew on the aircraft simply because of the anatomical damage. He understood the legal necessity for providing death certificates, as well as the importance to the peace of mind of surviving family members to provide a casket that contained a portion of their loved ones. The entire team came to understand the latter goal and worked tirelessly to achieve it.

The FBI and the armed forces provided fingerprints of those in their files. Families were asked to provide the names of the dentists used by the crash victims. The State Department worked with foreign governments to obtain data on victims from foreign countries. FBI agents fanned out to locate witnesses to the accident; others went to the homes and offices of some of the victims and collected fingerprints from items known to have been used recently by them. The witness statements and other data ruled out a bomb as the cause of the accident. DNA testing was ruled out as an identification method because of the absence of a reliable control sample for most of the remains. The FAA sent its own physician from the Civil Aeromedical Institute with two objectives. The remains of the pilots needed to be identified and properly secured to perform toxicology tests to rule out the use of alcohol or drugs.

Men and women in yellow suits ringed the water-filled main impact crater when the PWI crew began to pump the liquid into the large black tank truck. The only sound was the diesel engine and hum of the blower that produced the vacuum. Many expected to find most of the fuselage in this crater, as well as most of the passengers piled on top of one another like a deck of cards thrown haphazardly into a drawer. As the last of the water disappeared into the hose, everyone's attention was focused on the bottom of the hole. A few looked at each other in amazement; some glanced across the debris field. "They're all out there," an unidentified worker murmured to her companion, tears welling in her eyes. "My God, they're all out there."

The debris field extended nearly 700 feet to the east and northeast in a 60-degree arc. From the air, it looked as if someone had taken a handful of small stones and thrown them outward and downward, scattering them across the ground. The removal of the first remains from the site occurred on November 2, two days after the crash. Recovery workers began to exhibit signs of stress as the week progressed. As they crisscrossed the site, they began to associate items with the victims. Photographs of the victims were supplied by families, and when the workers came across recognizable clothing or jewelry, they associated the items with the face of a victim. The team worked quickly to remove all obvious body parts and unrecognizable masses of tissue from the site. As the ground continued to dry out, however, new discoveries were made under the soil surface wherever a slight discoloration remained wet, indicating more human remains.

People took various memories with them from the site work, but the events of the first two weeks of November 1994 undoubtedly changed the lives of everyone involved. Rick Clinton recovered the feet of Patrick Henry. Several workers reported hearing the sound of crying at times even though nobody could be seen crying. An oft-repeated statement of incredulity made by recovery workers was about the lack of blood, which had been washed away by the rain. When the tasks became almost unbearably grim, someone would find an intact Bible lying open to a passage, which they would read to the others. The Bibles always seemed to be open to a passage that comforted and encouraged the workers. Almost everyone I spoke with wondered at how the Bibles seemed to escape the destruction.

The task of clearing the site of remains took another seven days, until November 9, when American Airlines took control of the crash location. American contracted with PWI to continue the cleanup of the remaining wreckage and contaminated soil. Many of the workers continued to assist in the cleanup of the aircraft debris and personal effects after the bulk of the remains were removed.

Several days into the operation, Dr. Dennis visited the site. The morgue operations were progressing slowly; about a dozen identifications had been made. The remains of First Officer Gagliano had been identified, but not those of Captain Aguiar. Walking to the north of the impact crater, Dr. Dennis noticed a mass of piping and yellow and orange fabric that contained some numbers on a metal identification tag. The serial numbers were traced to the inertia reel of shoulder harness on the captain's seat. There was no indication that a body existed in the jumble of metal and fabric, but Dr. Dennis had the mass transported to the morgue, where it was carefully pulled apart in the presence of the FAA. Inside the captain's seat were the partial remains of Orlando Aguiar. A positive identification was made from a tattoo. With both pilots positively identified, samples of tissue were sent to the Toxicology and Accident Research Laboratory in Oklahoma City. Screening tests showed no indication of drug or alcohol use.

When the trailers arrived with remains at the morgue, National Guard soldiers had the job of locating the individual bags and bringing them to the forensic team inside the large main hall of the red brick armory building. Locating a specific bag inside a trailer that could contain more than 1000 small red plastic bags could take hours. To maintain the temperature in the refrigerated storage units, the doors needed to be closed. Temporary lighting was not bright enough to illuminate corners, so flashlights were often used to search. "Sometimes they were almost frozen when they came out," reported one of the people working at the site about the soldiers. "But those kids did it without question."

Jeff Davis represented American Airlines at the morgue. Davis had come to American from NASA, where he worked on another well-known disaster: recovering the remains of the crew of the space shuttle *Challenger*. The help and cooperation from the airline was graded as good by the managers of the morgue operation, and Jeff Davis respected the decisions of the coroner and his staff. Toward the end of the identification process, when American Airlines began pushing for caskets to be released so as not to delay their shipment schedule, the overworked and tired funeral directors at Remington held their ground and insisted they wait. As one said, "no casket was leaving here until every possible remain was properly processed and documented and the exterior of the coffin was cleaned and every fingerprint was removed" so the deceased could continue to be accorded the highest level of respect.

On the last day of the morgue operation, spirits were high because the team had identified all 68 people aboard the aircraft, and each family would receive a casket containing remains of their loved one. However, a sense of melancholy was also pervasive. The team had worked hard and felt they had come to know the victims. Even more, they had cried, commiserated, pulled each other along, and even laughed together during the three weeks of their association.

The previous day ended late, and the crew was tired. Members of the Indiana Funeral Directors Association worked frantically but carefully on the last day. The caskets containing the remains left the building to be shipped to the victims' homes. In addition to the 68 individual caskets, 19 common coffins containing unidentified remains were transported to the chosen site of a mass grave in Merrillville, Indiana. Cemetery workers had dug two 30-foot rows of graves in preparation for burial of the remains and had assembled burial vaults to hold the caskets in a far corner of the cemetery. The burial was unannounced; no one wanted to alert the media with the large assemblage of vaults and caskets arriving throughout the day. Lights had been set up in the event the burial continued past sunset.

Up until the time the morgue operations closed down, workers at the crash site continued to find remains and sent them to the armory in Remington. Remains that were found after the morgue closed were cremated.

National Guard soldiers stood by to disassemble and clean the morgue. Billets that had been set up to house the out-of-town workers through the project were removed; tents were taken down. Clerks continued to input data into the computers that tracked the remains and the progress of the work. The dental team had identified 34 individuals by the bits of dentition found at the site, which was estimated to be less than 9 percent of the total. Two portions of the lower mandible of one victim were found, one piece containing four teeth and another containing two teeth; they were recovered from two locations, 50 feet north of the main impact crater and 275 feet east. A passenger who had undergone open heart surgery was identified by the stainless steel wire around her sternum. Most of the remaining victims were identified by fingerprints. Sixty-eight death certificates all stated the same cause of death, "Multiple anatomical separations due to (or as a consequence of) high velocity impact due to (or as a consequence of) aircraft accident."

The embalming of the last remains was finished late in the day; the lid was closed and sealed on the final communal casket. The exhausted funeral directors decided to take the final caskets to Merrillville themselves. They went to change into suits and drove the last of the hearses north along highway 65, past the Roselawn exit, to the Calumet Park Cemetery. The shadows were long when they arrived. As the last casket was lowered into the ground, the professionals broke down and cried for people they had not known in life but had come to know in death.

One of the final acts at the crash site occurred in December. The large pile of contaminated soil was removed to a hazardous waste dump, reportedly near Anderson, Indiana, the hometown of the Stansberry family. Farmer Clarence Hanley insisted on two dump truckloads of new soil from a specific area of the state to replace each load removed. Work crews moved the remaining orange slop and graded the field, which continued to give up pieces of the airplane and body parts, necessitating two returns of PWI to do more cleanup work.

Mr. Hanley understands the way rocks migrate to the surface because of erosion and weathering. He also knows that airplane parts and bones will be surfacing for years. He is aware that the soybean field had become a burial ground. In the spring of 1995, he plowed and planted the hallowed earth. Around the impact crater and debris field, seed did not grow.

Passengers exit this ATR-72, parked at a gate in Indianapolis, from the rear cabin stairs. The front cargo door, located directly behind the cockpit, is open. Notice the scimitar-shaped propeller blades.

Captain Orlando Aguiar Aguiar family

First Officer Jeffrey Gagliano
Gagliano family

Flight attendant Sandi Modaff
Modaff family

Flight attendant Amanda Holberg
Holberg family

Jonathon, Patrick, and Patricia Henry Severin family

David Shellberg Shellberg family

Brad Stansberry Stansberry family

Barbara Tribble Tribble family

Some of the members of the Lincoln Township, Indiana, Volunteer Fire Department, which responded to the crash of Flight 4184. Cap Wann is seated, holding the mascot; John Knapp stands to his left. Lincoln Township Volunteer Fire Department

The horizontal tail of the ATR separated in flight. It lies to the west of the main impact craters (left side of photograph).

The impact craters and debris field from directly overhead

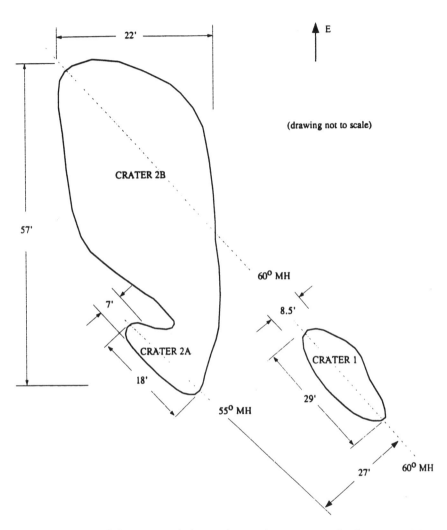

This drawing of the craters helps indicate their size and relative position.
NTSB

Recovery workers move among the stakes that mark the locations of human remains. Small body bags lie on the ground beside some of the stakes.

The tail cone section of the aircraft was one of the largest pieces of debris. The cockpit voice recorder and flight data recorder—the black boxes—were carried in the tail cone.

The tail cone shown from the opposite direction, surrounded by 8 acres of debris

This section of the rear cabin was the largest remaining piece of the passenger compartment.

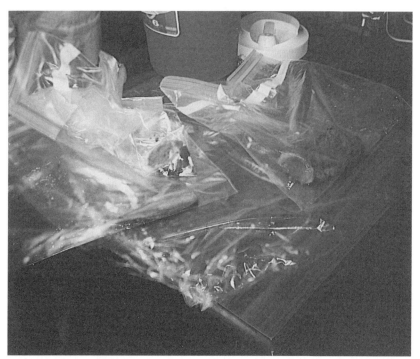

Tom and Terri Severin found these human remains in the field in Roselawn, Indiana, during the week-long NTSB hearings held in Indianapolis at the end of February 1995.

These 68 wooden crosses were built by Roselawn volunteers Helen Mudd, Chuck Mudd, and Julie Gunter for a roadside memorial service for the victims of Flight 4184 held the morning of August 18, 1995.

In 1984, Porter Perkins took this photograph of double-horn ice on unprotected areas of a wing. This formation of ice is the same type that accreted on the wings of the ATR-72 that Steve Fredrick attempted to land in Marquette, Michigan, in November 1993. He also saw this type of ice during his flights on October 31, 1994, several hours before the crash of Flight 4184. Porter Perkins and NASA

Ice formed by large droplets are shown here well aft of the protected leading edge of the wing. Porter Perkins and NASA

Jenny Stansberry, Dawna Aguiar, and Terri Severin stand at the memorial to the victims of Flight 4184 in Calumet Park Cemetery, Merrillville, Indiana. Each victim's name is etched on the black marble wall. The sundial pedestal in the center reads: "In Memoriam—American Eagle Flight 4184—October 31, 1994."

These two rows of graves, covered in pine boughs, accommodate the 19 caskets that contain the unidentifiable remains of the 68 victims of Flight 4184. Jennifer Stansberry

7

November 1994: A month of madness

Professional pilots pragmatically accept an aircraft crash and the loss of a pilot's life. This philosophical outlook is acquired during the years of climbing the aeronautical career ladder. Every flyer who has come up through the ranks of flight instructor, charter pilot, or night freight pilot, running checks, packages, or auto parts in ill-equipped, last-generation aircraft with unsympathetic bosses and unforgiving schedules has lost a friend or acquaintance. Many have nearly become statistics themselves, cheating death to fly again on another dark, stormy night. We fight hard to gain each rung of that ladder, hoping to land a job with a secure line flying the heavy iron.

Losing a fellow aviator who falls asleep and flies the aircraft into a mountain or tries to push a bad situation until it becomes untenable is part of the business. Losing two competent pilots, two flight attendants, and 64 passengers in a perfectly functional airliner is extremely disconcerting. We understand the ramifications of mechanical failure, the stupidity of disregarding aircraft or operational limitations, or the inanity of running out of fuel. Those are careless acts. When a modern aircraft unexpectedly flips on its back and plummets to earth with no chance of recovery, however, we grasp for any possible explanation, hoping to learn whatever we can to help ensure our own safety. So was the reaction to the loss of Flight 4184.

Almost imperceptibly, the tone of the pilot conversations in the crew lounge changed within days of the accident. We needed to put aside our initial discussions of sorrow, shock, and disbelief to stay out of a similar situation. It was not a lack of feeling—it was survival instinct. The topic focus transformed small group discussions on the growing suspicion of the cause and circumstances surrounding the accident into spirited debates regarding the new operational rules and the resurrection of the old tales of previous incidents. Those of us who had experienced control problems in ice were sought out for conferences with other pilots who had only had "normal, run-of-the-mill" encounters with ice accumulation and control anomalies—the ones we accepted as a part of life in the ATR.

Years ago, the National Advisory Committee for Aeronautics (NACA), the forerunner of the National Aeronautics and Space Administration (NASA), conducted research on a number of airfoil sections. The committee published data on the performance characteristics. At the time, the airline industry could not afford to do this research, and the shapes created by NACA have been used and modified for decades. ATR created a marvelous airfoil from a standard NACA design; it decreased drag while increasing lift, which made it more energy-efficient. Unfortunately, the same changes that made the airfoils such efficient producers of lift also created efficient collectors of structural ice, according to Porter Perkins, a retired icing expert who worked with both NACA and NASA.

Ice disrupts the clean flow of air over the wing and tail surfaces, and the results can be unpredictable. The first ATR icing incidents caused changes in the operational guidelines of aircraft, shrinking the performance envelope. Additional incidents occurred in rapid succession around the world and more than yearly at Simmons. The first ATR lost to ice crashed in the Alps in 1987. Patchwork fixes came and went, but the incidents still occurred with regularity. Operating airspeeds changed several times in the first years. By 1989, when Jeff Gagliano and I joined Simmons Airlines, a subsidiary of AMR Corporation, the envelope had shrunk further, and we were operating with new restrictions and higher speeds in ice. Airplanes buffeted, rolled, and exhibited other behavior not expected of a large regional airliner. Pilots got used to the mushy feeling of the ailerons during some operations, and we all lived by the credo "go fast and live."

Simmons Airlines operates in one of the most hostile icing environments in the world, and as a result of the procedures used by the air traffic control system at Chicago O'Hare, the airline's ATR aircraft are often operated continuously in ice from takeoff to landing. For years the company also operated Shorts 360 aircraft on the same routes without problem. The Shorts was forgiving in the ice, and many pilots refused to upgrade to the larger ATR, even though the pay was higher, specifically to avoid the negative reputation of the ATR. The seasoned pilot group had a respect for the ATR in ice that bordered on paranoia. Then we lost one. Flight 4184 was not a near-miss with recovery after 20 seconds of terror. Two of our pilots lay in a soybean field. The game had changed.

Classifying the level of severity of icing encounters in an aircraft is a highly subjective, aircraft-specific determination by the pilot. A minimal inconvenience to a large jet airliner can be a life-or-death scenario for a small twin-engine aircraft, even though both might be certificated by the FAA for flight into ice. Reporting such conditions to flight service stations or air traffic control is recommended as good airmanship, but it is not mandatory. An absence of pilot reports does not indicate there is no ice, and pilots are trained to determine the type of aircraft making reports to gauge its danger for their airplane.

As with turbulence, aircraft structural icing comes in many shapes and severities, and each type is defined in the *Aeronautical Information Manual,* or AIM. In general terms, three types of ice exist. *Rime ice* is found mostly in stratified clouds or drizzle and is identified by its cloudy, frostlike, irregular-shaped, rough-surfaced formation; its major effect on airfoils is spoiling the smooth airflow, decreasing the aerodynamic efficiency. *Clear ice* is found mostly in rain or cumulus clouds. It forms in the presence of large drops of water that are near or below freezing while still in a liquid state. When these drops hit the aircraft, they flow out and freeze, forming a clear, hard, heavy glaze accretion that affects the aircraft by adding significant weight. If the clear ice formation is rough, it can affect boundary layer airflow and significantly reduce the lift capability of the wing. *Mixed ice* is a combination of rime and clear ice. Depending on the liquid water content and size of the droplet, mixed ice can resemble clear or rime ice. On occasion, as a result of its rapid accretion, irregular shape, and high weight, mixed ice can be extremely dangerous.

The severity of the ice is classified by four definitions. *Trace icing* means the accretion sublimates, or turns back to vapor directly from its solid state, at a slightly slower rate than its development on the airfoil. This exposure is generally not a problem for most aircraft. *Light icing* is a level of development that can be disposed of by the deicing equipment and presents no significant hazard to aircraft with deicing equipment. *Moderate icing* is a rate of accumulation that might become hazardous after short encounters, and the use of deicing and anti-icing systems or diversion is necessary. *Severe icing* requires the immediate diversion of the flight because of the inability of the deicing equipment to cope with the accumulation. The problem with the reporting and classification systems for icing is that the qualifiers are subjectively interpreted; no widespread, globally recognized, objective system or quantitative data-gathering technology exists for reporting icing encounters and environments.

˙st time I picked up ice on a single-engine Bonanza, a six-
˥e without any deicing equipment, I was descending
‿ı a thin stratus cloud layer to warmer temperatures, and I en-
‿ountered what I now consider to be light rime ice. The ice quickly
melted 1000 feet below the cloud, but at the time, I was watching that
ice like a hawk. Had anyone bothered to ask, I would have said that
the icing was moderate or severe, but experience tempers and tinges
a pilot's observations.

Any ice accretion on an aircraft, even frost, can severely affect
performance, and a prudent pilot always monitors the situation carefully. Part of that monitoring is the application of judgment based on
experience. Most pilots downgrade the severity of their ice encounter
assessments over time as they become more comfortable in ice conditions and develop a "seen that before and it was OK" attitude. With
any aircraft, and especially the ATR, in my opinion, this attitude can
be fatal.

Every aircraft must adhere to the same physical principles and
overcome the same gravity and drag with the use of lift and thrust.
The variables are numerous and the technical details are complex,
but the underlying fundamentals of controlled flight are basic and do
not change.

As an airfoil, such as a wing or horizontal tail, moves through the
atmosphere, the air in its path is moved above or below the wing. As
a result of the curvature of the wing profile, air that passes over the
wing surface travels farther than the air that passes under the wing.
At takeoff or flight speeds, the pressure of the air on the top of the
wing decreases with increased speed, and the pressure below the
wing increases, creating lift.

As a result of lift, there is a consequence called drag. *Induced
drag* differs from the parasite drag that develops because of form resistance and skin friction. Hold your hand out of the window of a
moving car and feel the pressure of the wind. If you hold your hand
horizontally like an airplane wing and you tilt the front of your hand
up slightly, you can feel a force pushing your hand up and to the rear.
This illustrates both lift, which pushes up, and induced drag, which
pushes to the rear. Hold your hand vertically with the palm facing flat
into the airflow, and you feel a strong force pushing rearward. Because no lift is generated by this position, your hand is now affected
entirely by parasite drag. Some of both types of drag exist in the airfoil example, but human arms are not sensitive enough to measure
the difference, nor is the hand a perfect airfoil.

The lift produced by a wing overcomes gravity, while the jet or
propeller engine overcomes drag, and the aircraft flies. The main wing

lifts the aircraft into the air. The horizontal tail trims the airplane attitude in pitch by producing lift in an upward or downward direction using the elevator surface control operated by the pilot or autopilot.

The angle at which the airfoil strikes the air relative to the movement of the aircraft is the angle of attack. The angle of attack consists of two lines. Draw an imaginary line from the forward tip of the wing to the rear tip of the wing; this is the chord line. The other line comes from the direction the aircraft moves through the air. When an airplane takes off, for example, the pilots raise its nose, which tilts the chord line to form a greater angle of attack, which adds lift to the aircraft at the low speed needed for takeoff. During cruise or level flight, the airplane weighs about the same amount, and the wing must produce lift to keep it flying. However, because of the increased speed, the angle needed to produce the lift is less; therefore, the nose of the aircraft seems almost level.

To make an airplane perform better at low airspeeds, flaps at the rear, and sometimes at the front, of the wing can be deployed, forming a new chord line drawn from the tip of the lowered flap to the forward tip of the wing. When the pilots of Flight 4184 placed flaps at 15 degrees for level flight, the crew and passengers experienced a lower and more comfortable deck angle because of the increased angle of attack formed by a change to the chord line. Measurements along the chord line, which is straight and more easily defined than a line along the curved airfoil, are always given in percentage of chord, sometimes referred to as percentage of mean aerodynamic chord, starting from the front of the airfoil. For instance, 10 percent of chord with a total chord line length of 100 inches means a point 10 inches behind the leading edge of the wing.

Every wing has a critical angle of attack. When that angle is reached, the airfoil will stall, losing its ability to lift, because of disrupted airflow over the top of the wing. On a clean wing, the results are predictable and do not change. The wing loses its lift at the same angle of attack, regardless of speed, every time. When contamination, such as ice, builds on a wing, the formation constantly changes, and the changes in the airflow are unpredictable. The wing might therefore stall at a lower angle of attack or show other anomalous effects. As a result of the unpredictability of ice on wings, aircraft designers try to create a wing that either does not allow the formation of ice or sheds it using deicing equipment, thereby maintaining the original shape of the wing. Contaminants affect airflow, and pilots hope they can leave ice before something happens. Problems arise when the pilots cannot see ice on an unprotected surface, such as the top of an ATR wing in dark conditions.

The early details of the accident investigation were not widely disseminated to Simmons pilots, and, from our perspective, the airline flight management department seemed to be in a panic as a result of the need to maintain operations while finally facing the poor performance history of the ATR in ice. Senior captains were called into the office for what we dubbed "cheerleading sessions." Ed Harvey, the Simmons Airlines director of flying, Ed Simon, the domicile chief pilot, or other lower-level administrators requested cooperation in maintaining a positive attitude among the pilots. Few signed on to the company line, however, and morale deteriorated.

Most pilots understood the accident was a result of a loss of control after ice accretion almost from the time of the first reports. However, the speculation over the exact scenario was widespread. Management's apparent lack of honesty hurt the airline in the long run. One story widely retold among the flight crews involved a conversation between Ed Harvey and Captain Dave Harthorn.

Dave Harthorn is a senior pilot at the airline and has been flying the ATR almost since its introduction to the company back in 1986. He had seen previous incident investigations and wanted to know the bottom line on this accident, so he approached his longtime friend Ed Harvey and asked, "What happened?"

Ed Harvey looked straight into his eyes and said, "Dave, you have my word that it wasn't ice, and the aircraft didn't stall."

The ATR-72 flown by Captain Aguiar and First Officer Gagliano did not stall that day, according to all reports and decoded flight data. However, the NTSB suspected ice within two days of the crash. Since Ed Harvey was involved in the investigation, he was privy to the NTSB's reports, and I would be surprised if he did not know that it was ice when he spoke with Dave Harthorn.

Within a couple of days of the accident, management began implementing operational changes, reminded pilots of long-standing and well-understood guidelines, and strategically reworked some procedures. The first operational change was a prohibition against the use of the autopilot when operating in icing conditions. For long-time ATR pilots, this was almost a verbatim repeat of steps taken after the 1988 incident near Wausau, Wisconsin, when another ATR was almost lost to ice.

Removing the autopilot from the control of the aircraft seemed like a logical first step in 1989. In 1994, with incident counts mounting since 1986, including at least one fatal accident, we had seen this before. Previously, we had been assured the problems had been resolved, and no more incidents would occur if we followed the rules. ATR pilots at other airlines had been told the same thing, but they, like the Simmons pilots, had not been told many details about the re-curring problems of the other operators around the world. We had not been sufficiently informed regarding other loss-of-control inci-dents, including incidents in 1991 over Great Britain, in 1993 near Newark, New Jersey, in 1993 near Quincy, Illinois, and in 1994 near Burlington, Vermont.

The logic of precluding the use of the autopilot in icing condi-tions was based not on the belief the autopilot had anything to do with the development of the aerodynamic problems, but that its use masked the early warning signs of the impending loss of control. From the perspective of what we knew in November 1994, it was lu-dicrous; from the evidence mounted since then, it bordered on gross irresponsibility.

The ATR control system uses standard cables and manual force to move the control surfaces. Unlike jetliners, ATR models have no hy-draulically boosted primary flight controls. To remove some of the control wheel forces from the pilot, designers included surfaces that extend forward of the control-surface hinge line, called *aerodynamic balance horns*. When the control is moved against the airstream, be-cause of the location of the horns, it is affected by a negatively cor-responding airflow that reduces the control pressures felt by the pilot. Should the air pressure on the control surface be removed or reduced for any reason, such as a detached or disrupted airflow caused by ice, the effect of the horns would require pilots to place force opposite to what would be expected to move the controls. This phenomenon is known as *hinge moment reversal* and is an undesirable characteristic for an aircraft. Normally, if a pilot wants the aircraft to turn to the right, he or she induces a control force through the wheel moving the ailerons, and the aircraft turns. When hinge moment reversal occurs, the pilot feels the force turning the wheel without his or her input and has to hold back the wheel against the pressure by applying strength in the opposite direction to what is expected. When that happens, it tends to occur with a rapid onset. A professional test pi-lot who knows the anomaly is coming has his or her hands full in controlling the aircraft, but an airline pilot flying in cloud who has never experienced the situation before is in over his or her head.

Essentially, what we were told to do by hand-flying the airplane in any icing conditions was to act as test pilots with passengers on board so that the airline could continue to generate revenues. In De-cember 1986, after 50 percent of the Simmons Airlines fleet of ATRs encountered an ice-induced loss of control within an hour on the same day approaching Detroit, the FAA quickly grounded the entire fleet in icing conditions. At the time of this safety move, fewer than

20 ATRs were operating in the United States; the 175 ATRs operated by U.S. carriers in 1994, representing an estimated $2.5 billion investment to the airlines, constituted a significant lobbying force in Washington, D.C. Grounding the aircraft did not make financial or political sense.

In the same memorandum, which appeared on each of our flight releases after the accident, we were reminded of practices not in place when the company equivocally directed, "As stated in ATR Aircraft Operating Manual Volume 1, 'Holding procedures must be accomplished with flaps zero and at an airspeed of not less than VmHB0 icing and preferably at a speed equal to or greater than conservative maneuvering speed for the ATR-42 or 175 knots for the ATR-72.'" VmHB0 is the minimum speed in a high-bank turn with the flaps at 0 degrees.

If this sounds confusing to the layperson, it should. Pilots were also confused, and we greeted the memorandum with "What the hell does this mean?" Many looked up the passages in their manuals that related to holding procedures and speeds. Some pilots took the memo literally and began operating their aircraft in a manner that went against conventional wisdom and experience. Others stuck to the long-standing practice of lowering the flaps to 15 degrees whenever maneuvering in icing below 185 knots. Debates sprung up between pilots who believed a causal relationship existed between ice and the accident and others who did not.

Although it is true that the philosophy of placing the flaps at 15 degrees while holding is not mentioned as a procedure in the operations manuals, neither is the practice prohibited. The training department is split on the use of flaps below a certain speed. Almost all Simmons instructors with the most experience in the aircraft choose to set flaps immediately to 15 degrees on slowing below 185 knots in the ATR-72; they chastise those who do not. Some instructors who are used to operating the craft in Nashville or Miami are not so demanding, deferring to the pilots' discretion. One thing clearly stated in the company manual is "ATC holding pattern speed limits for turboprops of 175 KIAS [knots indicated airspeed] must be observed." Eagle Flight 4184 was in full compliance of this American Eagle policy, which carries the threat of discharge if violated.

The options were clear, felt many of the pilots. With a heavy ATR-72 in icing conditions and turbulence during a hold, most would do the same as the ill-fated crew had done; some held to the belief they would be safe at 175 knots with zero flaps and a high deck angle. Now that what we would have done in the past had proven to be unsafe, most pilots set out to refuse holding patterns and extended periods of operation in the ice, and many informally agreed to a new minimum speed in ice of 210 knots. This speed was 16 percent higher than ATR had previously stated as a conservative speed, and it made some pilots feel better. Unfortunately, they eventually had to slow to between 210 and 180 to get the flaps to 15 degrees anyway, as the computer would not allow the flaps to move at airspeeds higher than 185 knots.

Our faith in the credo, "go fast and live," had been shattered. The other consensus conclusion that pilots drew from the initial reminders was that, if something happened, the company was padding its own defense by providing hooks from which to hang pilots. Captains, especially some senior captains, took offense to the new rules of operation and fought back by asking for specific guidelines. None were provided, other than referencing "approved company procedures." Did that mean what was written or what was trained? Financial pressure kept pilots flying, but soon people began throwing things and slamming doors in the crew lounge.

On November 4, the following message was transmitted from Ed Harvey to all of the pilots via e-mail.

It is imperative that all pilots landing at Marquette County Airport under emergency conditions notify the weather service station of their emergency at least 10 minutes before making the emergency landing at Marquette. This is necessary to ensure that the proper crash/fire/rescue equipment is on the scene during the touchdown portion of the landing. In order for the airport CFR unit and the Negaunee Township Fire Department to give you the best service possible, this 10 minute notification is required, time permitting. There is nothing more dissatisfying than for the fire crew to show up at the emergency location after the emergency has been terminated. Thank you.

What was this? Had somebody had an emergency landing in Marquette? Was it ice-related? Nobody knew, but in our growing suspicions, the bulletin loomed large.

On November 11, we received the following message from Ed Harvey:

Although the desire to learn the cause of the accident is of great importance to all of us, and in spite of the fact the NTSB has access to a significant amount of data related to the accident, as of this writing, neither the NTSB, FAA, ATR, or the Company yet know exactly what happened. The NTSB group assigned to the accident is currently working to confirm or eliminate potential causes which they have identified to date. The focus of their studies has moved to the NTSB Performance

*Group and their evaluation of the Flight Data Recorder,
which is being accomplished in France by ATR. In conjunction with the investigation, the FAA has also formed a group
to review, in conjunction with the DGAC, any certification
data on the ATR aircraft which may be pertinent. We estimate
that it will take approximately two months before any results
from their efforts will be available. The FAA is expected to issue guidelines on the ATR operations next week; however, with
the implementation of the following procedures, we expect no
further procedural modification as a result. We also expect
the FAA to issue next week temporary guidelines to ATC nationwide asking for their assistance in accommodating requests from ATR crews for modifications to routings and/or
altitudes to the extent necessary "as practical" to minimize
aircraft exposure in ice accretion conditions. We will do our
best to keep flight crewmembers apprised of the status of the
investigation as information becomes available.*

In addition, we received a number of additional operating procedures that were not new or definitive. Pilots were still no closer to the
truth on the airplane. The blame for any further incidents was being
laid at the feet of the pilots; the company always left itself a legal or
technical loophole in its policies. The wording and selection of these
rules of operation also served to add to our tension.

"When icing conditions are reported or forecast for the planned
route of flight, and ATC advises that an expected delay or hold may
be taken either in the air or on the ground, the delay must be taken
on the ground." Rarely does air traffic control release a flight with an
anticipated hold in the air like they did with Eagle Flight 4184. Were
we supposed to ask if there was a hold anticipated down the line? If
there was, did we sit on the ground because of what might happen?
Is an unexpected hold dangerous? If so, why?

"Dispatchers will monitor atmospheric conditions and pilot reports for actual and/or forecast icing conditions and modify flight
plan routes and/or altitudes to the extent practical to minimize icing
conditions." The use of "to the extent practical" bothered almost
everybody. Was it practical to have a flight to South Bend 60 miles
east of Chicago climb to 11,000 to get on top of a cloud deck when
normally we only climb to 7000 feet on most days because of other
traffic? If we told the tower controller before takeoff we needed an
unlimited climb to 11,000, we could run out of fuel sitting in the holding pad on the ground while air traffic control waited for an opening
in the flow of traffic. If we told the controller after departure we
needed the climb, we could be stuck at 5000 feet for miles while jets

climbed above us. It happened and still does. The dispatcher's job is to supply pilots with the best, safest, and most efficient routing by monitoring weather and pilot reports. This is new? Most pilots believed this memorandum was a waste of paper.

"When an ATC clearance for a routing other than filed is received, the flight crew will review available weather reports, forecasts, and pilot reports to ensure the new routing does not 'significantly' increase the flight's exposure to icing conditions from that initially filed." The qualifier "significantly" appeared with quotation marks in the memo without any further definition. If we anticipated 30 minutes of flight in icing conditions, did not pick up any ice, but were still in icing conditions and the controller added another turn that added another 10 minutes, or 30 percent, was that significant?

Pilots make decisions based on what is occurring to their aircraft. No pilot will sit in weather in which he or she does not feel comfortable just to avoid a discussion over whether the change in plan was significant. Simmons added the prohibition on the use of the autopilot in moderate or greater turbulence. The gauging of turbulence is as subjective as ice accretion. Why the restriction? We were not told the reason, which added to frustration and conjecture. It seemed obvious the company knew more than it was telling us. Our greatest concern was that we would never know until a repeat of the accident occurred.

Our answer came when somebody posted a letter from the NTSB to the FAA within a day of the distribution of the memorandum. It was a six-page, single-spaced letter to David Hinson, administrator of the FAA, from Jim Hall, the chairman of the NTSB. It was dated November 7, four days prior to Ed Harvey's memo. After a short introduction and fact review, Chairman Hall jumped to the heart of the matter.

The investigation of the accident is continuing, and probable causes have not been determined. However, based on evidence uncovered in the investigation, the National Transportation Safety Board believes that the Federal Aviation Administration (FAA) should take immediate action pertaining to the ATR-42 and ATR-72 aircraft. Evidence from air traffic control (ATC) sources and the airplane's flight recorders have prompted the Safety Board's concern that the loss of control leading to the steep dive might be attributed to the weather conditions encountered by the flight and the characteristics of the aerodynamic design and flight control systems of the airplane. The Safety Board is aware that similar uncommanded autopilot disengagements and uncommanded lateral excursions have occurred on the ATR-42 airplanes during the past 6 years, although none have re-

sulted in a nonrecoverable loss of control. The Safety Board investigated one such event that occurred on December 22, 1988, at Mosinee, Wisconsin. A review of the FDR data for that airplane showed similarities with the data from Flight 4184. That is, as the angle of attack increased, an autopilot disengagement and rapid roll command was evident. In the 1988 occurrence, the flight crew regained control after losing 600 feet of altitude, and the subsequent landing was uneventful. On June 28, 1989, ATR issued Service Bulletin ATR-42-57-0018, Revision 1, which described the installation of vortex generators on the upper surface of the wing forward of the ailerons. The FAA subsequently issued an amendment to AD 89-09-05 which accepted the compliance with the ATR Service Bulletin as terminating action for the AFM limitation regarding use of the autopilot when operating in icing conditions. The ATR-72 incorporated the installation of vortex generators in the original design. Thus, there was no prohibition against use of the autopilot when operating an ATR-72 in icing conditions. Although the Safety Board was not involved in the aerodynamic performance analysis or flight test activities that led to the ATR Service Bulletin or FAA AD, the Board believes that the vortex generators were intended to (1) prevent premature boundary layer separation on the aileron control surface(s) as a result of an in-flight accumulation of ice on the upper wing surface aft of the de-icing boot; and (2) to assist in the recovery from a roll departure. Nevertheless, the Safety Board is aware of another occurrence wherein the flight crew of an ATR-42 experienced roll control difficulties while operating in icing conditions, although the vortex generators had been installed. Therefore, the Safety Board is concerned that an amount of ice that can be accumulated under some flight conditions encountered during winter line operations may be more critical to the flying quality of ATR-42 and ATR-72 airplanes than to other airplanes. The Safety Board believes that a slight amount of ice accumulated under some conditions may produce boundary layer separation on one or both ailerons that can result in abrupt changes to the aileron hinge moment. The Safety Board believes that the control wheel force-versus-airplane rolling moment characteristic may be unstable depending upon the angle of attack and the magnitude of aileron deflection. The circumstances of this accident and the previous incidents involving ATR-42 airplanes indicate that the use of the autopilot can mask the onset of the lateral control instability. The Safety Board believes that the autopilot, operating in a lateral navigation or altitude hold mode, will provide commands to the lateral control system that compensate for aileron hinge moment changes until a given force or position error threshold is reached. The autopilot will then disengage, and the lateral control system will react to the abnormal aileron forces. Because the FDR does not record control wheel force, the Safety Board could not determine the amount of pilot force needed to counter the uncommanded aileron deflection. However, the Safety Board believes that it is likely that both pilots were attempting to exert the force necessary to level the wings and were unable to do so, except for those instances in which the angle of attack was lowered sufficiently.

Pilots read and recopied the letter, which contained data not previously known other than by rumor. However, the correlation of the accident to the 1988 incident caused the most commotion. We had been assured that the problem that caused the 1988 incident had been fixed. The wording from the 1989 FAA airworthiness directive seemed to be rehashed verbatim for 1994. The caution we had overlooked as nonsensical rhetoric was not initially relayed in its entirety, which would have given us more concern. "Pilots should be advised that prolonged operations in temperatures near freezing with visible moisture should be avoided. Operations in these conditions, or with visible ice on the aircraft, may result in asymmetric wing lift and associated increased aileron forces necessary to maintain coordinated flight. Whenever the aircraft exhibits buffet onset, uncommanded roll, or unusual control wheel forces, immediately reduce the angle of attack and avoid excessive maneuvering."

Four days before Ed Harvey and American Eagle tried to assuage us by sending memorandums and stating the NTSB knew no facts surrounding the accident, they appeared to have the information but were not telling the pilots. Jim Hall also recommended five urgent actions be adopted by the FAA to remove the ATR from hazardous conditions and protect the flying and general public. The next ATR might not come down in an empty field, but in downtown Chicago.

Pilots understood that the suggested actions were not submitted by the NTSB without significant data; we also understood the minimal chances of the FAA acting without outside pressure. We believed we were not going to get any help from David Hinson's agency because Hinson was the former chairman of the board for the defunct Midway Airlines; previous FAA rulings under his administration had been favorable to the airlines. We did, however, hold a glimmer of hope.

A turning point took place several days later when we were confronted with further modifications to the operating procedures. The release message that had begun the month as a single paragraph now filled almost an entire page with provisions and caveats. The FAA issued further revisions on November 15, and we were told our November 11 conditions already complied with the new governmental mandates. A frustrated senior ATR captain told me, "I understand they have to keep the airplanes flying or else we are out of business, but I am not going to get killed over it. I have already decided that if I don't like what I see, I am out of here."

By the week prior to Thanksgiving, things had been changing on a daily basis. Every time people reported for a trip after a couple of days off, the first question out of their mouths seemed to be "What happened since I left?"

On November 18, things changed on every flight. I flew four legs that day and had three different ATR operating-policy messages. We were conducting flights under conditional language, and new postings informed us of the "suggestion" that we only turn the aircraft at half the normal bank angle when we were operating in icing conditions. That one issue in itself indicated how hard the operators were working to keep the aircraft flying in growing evidence of a problem. At Chicago O'Hare, we routinely use parallel instrument approaches to land, and the controllers vector the flights onto the final approach course, often at the last minute. If pilots do not turn when instructed, it is easy to fly right through the localizer course. The controllers anticipate pilots will turn at a standard rate of 3 degrees per second, which yields a 90-degree turn in the direction of flight in 30 seconds. If the pilot turned at half rate, that same turn would take 60 seconds, and the aircraft could fly across its final approach course and into the path of another airliner. Most pilots ignored the suggestion and adhered to the normal rates of turning as the less-dangerous option, especially in the crowded skies over Chicago.

By the end of the day on November 18, Simmons had reached the bottom of the barrel of ideas when it issued the third change to the procedures. In addition to minimizing the bank angle to 15 degrees in icing conditions, when practical, we were to watch for "large, supercooled raindrops." As opposed to normal-sized, supercooled raindrops, we joked, but it wasn't funny. Flight management was telling us to be aware of encountering clouds with droplet sizes in the range of 40 to 200 microns. "How am I supposed to gauge a droplet size at 300 miles per hour?" was more than one pilot's response. The company also issued a flap-management warning stating "do not raise the flaps beyond 15 degrees until the airframe is clear of ice."

Again, we were not told why. We were carrying passengers aboard an airplane that was obviously unsuited for its duty, and we were the guinea pigs.

On the morning of November 19, the weather, which had been uneventful for most of the month, turned overcast. The thick stratus clouds formed without a lot of moisture, so icing was not a great concern, although at this point nobody was sure anymore where the dividing line between "no concern" and "concern" was. The approach to Chicago O'Hare put my plane behind a Boeing 757, and I had a mental picture about how our flight path would correspond with the larger aircraft. I was flying the full ATR-72 by hand from Fort Wayne, Indiana, and I knew we most likely would encounter a bump caused by the wake turbulence from the big Boeing and discussed staying a little high on the glideslope to avoid the worst of it. The deicing systems were operating normally, and we were in the cloud about seven miles from the airport and talking to the tower controller when we felt the first burbles of the jet's wake.

Captain Tim Riley, an ex-Shorts 360 pilot who moved to the ATR when our final Shorts aircraft was phased out of operation earlier in the year, sat to my left and glanced over at me. "I knew that we'd run into some of this garbage," he said. He had barely finished when we flew into the cyclone generated by the wings of the 757, and the heavy ATR-72 rolled rapidly to the right. The Roselawn scenario was still fresh in everyone's mind, and even though we only reached about 30 degrees of bank before I regained control, the roll got everyone's attention. "Eagle 393 is going to need the approach landline number after arrival because of wake turbulence," I tersely said over the radio even though I also was flying the airplane. "Roger." At 700 feet, just as we were about to break out of the clouds, the controller called back and gave us the phone number. I had ticked him off with the request—he could have passed it to the ground controller to relay after we landed and left the runway, but he was sending a message. After we parked at the gate, one of our flight attendants came forward and said, "What was that?"

"Wake turbulence," I said. "We knew the controller was setting us up for it but couldn't do much about it."

"Well you should have seen the people in the back," he commented with a nervous laugh. "I was facing them in the forward jumpseat and when we hit that roll, all I could see was 130 bugging eyeballs and 65 gaping mouths looking at me. That included the other FA."

"Too bad you couldn't see us," I said. "I think we have to wipe the eyeball spots off the inside of the windshield." We knew the roll was not related to ice, but the level of stress in the cockpits had hit the redline.

As Thanksgiving Day approached, the weather turned sour in the midwestern United States, and pilots' emotions boiled over. Thanksgiving weekend is always the busiest travel time of the year, and the terminals were packed. We were pressured to complete our schedules. Many of us did not think the friction between the pilots and management could get worse. Then the story circulated in the crew lounge about several pilots who refused to fly being referred to psychological counseling through the employee assistance program.

The captain of a commercial airliner is charged with the complete safety of the flight, and his or her judgment is supposed to be the final word under federal regulation. Airline management should respect the captain's decision. Unfortunately, Lance McDonald, vice president of flight operations for Simmons, must have seen it differently. In a press statement, McDonald expressed his feeling that the affected pilots were under stress attributable to the Indiana accident. What he did not state, and what the press was not savvy enough to inquire about, was the underlying career threat in that seemingly benign counseling suggestion.

Each pilot must pass an FAA physical examination every six months to act as pilot in command of an airliner, and with the examination, the pilot must fill out a questionnaire regarding any medical services, including psychological counseling, received since the last examination. Because the emotional stability of a pilot is always of concern to a future employer, the presence of any counseling on past medical certificate applications can prevent a pilot from moving up the ladder. By ordering psychological counseling, the company made an implied threat to the careers of the pilots, and most pilots were irate. We received no help from our union.

On the day after Thanksgiving, leaflets began appearing in the terminals, bookstores, phone booths, and even the bathrooms of Chicago O'Hare warning passengers about the dangers of the ATR in icing conditions. These leaflets were written by parties knowledgeable about the history of the ATR and Simmons Airlines, but nobody in the crew lounge admitted to having any part in their publication. I was one of the people who helped distribute them. I was flying a four-day trip during this time and will never forget how damaging one of those pieces of paper looked stuck in the *Chicago Tribune* in a store located in the G concourse. The distribution of the leaflets was an attempt to publicize our dilemma and was the catalyst toward notification of the public of a deeper problem at Simmons. The paper was two-sided, and text leaped off the front with the following large print headline and front page text:

WARNING!
American Eagle Passengers
Despite all of our warnings and past experiences, 68 people died needlessly on October 31, 1994, in the crash of American Eagle Flight 4184 in a soybean field between Indianapolis and Chicago. This accident did NOT need to happen and the FAA, American Eagle management, the Air Line Pilot's Association (ALPA), and Aerospatiale (manufacturer of the ATR aircraft you will be flying in on your trip today) could have prevented it.

American Eagle (Simmons Airlines) has had between 8 and 12 REPORTED incidents (and many more unreported incidents) relating to the LOSS OF CONTROL or IMMINENT LOSS OF CONTROL due to ICE on wings and tail surfaces over the past 8 years. Only luck saved these aircraft, several of which were recovered by the flight crew just SEVERAL HUNDRED FEET ABOVE THE GROUND.

The pilots of these aircraft have pleaded with the company and FAA to review the certification of these aircraft in light of each and every ICE incident that occurred over the years. MONEY has been the motivating factor in denying any indepth investigation, and now 68 people have paid the ultimate price. The pilot group of this airline are irate and we need your help!

Please review the information on the reverse side of this flyer and help us pass this along to the press, attorneys, and Congress. It may take Congressional hearings to make these aircraft safe during icing conditions. The FAA, ALPA, and the company have successfully swept past incidents under the carpet in the interest of PROFIT. Much more information is available! We, the American Eagle ATR Pilots, are highly trained professionals and will do all we can to make your flight a pleasant and safe experience. If the weather is clear this winter, sit back and relax because this is a good aircraft. If the weather is cloudy, snowy, or cool and rainy, think about alternate transportation methods because the majority of us will refuse to fly these aircraft until the icing problems are fixed. Even with the threats of management hanging over our heads, we will exercise our rights as the "ultimate authority" of the flight and protect your safety. Thank you.

The leaflets caught everyone off guard, and quotations from the text found their way into newspapers across the country. Simmons management ordered some employees to walk the terminals, quietly pick up any leaflets found, and watch for anyone distributing them. Pilots secretly relished the appearance and potential of these leaflets and groups gathered in the crew lounge to read the document between flights. Although this medicine was not for everyone, most pilots knew it would draw needed attention. The timing of the leaflet was fortuitous, because two days later, the ice hit the fan in Chicago.

On Sunday, November 27, the weather was reminiscent of that on October 31, with moderate rain showers, widespread low-level clouds, gusty winds, 40-degree ground temperatures, and a freezing level indicating a temperature inversion aloft. Tim and I knew this was bad ATR icing weather before we walked out of the motel in Champaign, Illinois, to begin our day. We talked about the nasty weather and our options during the long van ride to the airport. Tim pulled the flight release from the operations area, and I walked out to do the preflight check on the aircraft, taking an umbrella from the storage rack before walking into the cold, wind-driven rain. I was not happy; neither was Tim. This was the fourth day of a four-day trip, and while in the motel, we had both watched the Weather Channel, which reported a large area of this type of weather covering most of the upper Midwest. There would be no relief from the conditions today. After the preflight check, I walked back inside to check to see what Tim had found in the weather package.

"Moderate rain and some embedded convective activity," he reported with a grimace.

"Any ice reports?"

"It's early, but there are a couple that are already two hours old."

"What do you think?"

"I was just about to ask you," he laughed.

"Why don't we load up, see what the delays might be to O'Hare and check the radar from the ground," I suggested. "We can get the latest from dispatch at the end of the runway. If either of us don't feel right about it, we cancel and sit the day here. Agreed?"

"Fine by me." Forty passengers boarded the ATR-72, and we told the tower we wanted to spend some time at the end of the runway, at the same time requesting any pilot reports from the area. It was Tim's leg to fly, and I intended to keep us out of trouble. The radar painted light-to-heavy precipitation all around, and we picked a preliminary direction of flight out of the airport area and asked for an en route altitude of 4000 feet to Chicago O'Hare to stay below the freezing level. When Tim parked the airplane on the runup pad, I called dispatch on the radio for a weather update.

"Aircraft calling dispatch, go ahead."

"Good morning," I said, "It's 4395 in Champaign to O'Hare, and we want an update on the weather to the north." The dispatcher relayed current O'Hare and en route weather, and I asked for pilot reports, specifically those relating to ice.

"Well, I have several for moderate ice along your route of flight. One a Saab 340 at 14,000 feet and one a Beech King Air at 10,000 feet. I also have a report of freezing rain east of Rockford, aircraft type unknown, and another for light freezing rain 10 miles west of O'Hare."

I looked at Tim with incredulity, and he looked back the same way. Before the accident with Flight 4184, company policy precluded dispatch into freezing rain when there were two independent reports in the same general area. After the crash, the company had said "no dispatch into freezing rain or freezing drizzle, period." Yet, this woman was sending us into the teeth of the reports.

"Dispatch, 4395," I said, with as much control as possible. "Are you comfortable with releasing this flight from Champaign to O'Hare with two reports of freezing rain, one within the hour and 10 miles from our destination?"

A five-minute lapse occurred before her response, and I had to call her back to get it.

"4395, Dispatch," she said slowly. "I will never be completely comfortable dispatching a flight again. But the weather was west of your destination airport. Besides, it's only light freezing rain."

"That doesn't answer my question," I replied more tersely. "I know you are under pressure from management to dispatch every possible flight, but I want you to tell me that you are personally sanctioning this flight and that you feel fine with that decision."

The next response took seven minutes and was to the point: "4395, it's your call."

Tim and I looked at each other, incredulous. Each of us read the confusion and disgust in the other's face. One of the company rules most highly stressed in our operational training was designed to reiterate Federal Aviation Regulation's Part 121 requirement that dispatchers and captains be in *total agreement* as to the operation of a flight and that both parties must actively resolve any operational differences prior to the departure of the aircraft.

By abrogating her authority and turning the decision totally over to us, the dispatcher not only broke the company rule, but, in my opinion, violated FAR 121.533(b) regarding the requirement for a dispatcher to provide operational control of our airliner, and FAR 121.593, which specifically details a dispatcher's authority and stipulated the exigency of an *explicit dispatcher authorization* for our domestic air carrier flight to operate. In addition, I believe she violated FAR 121.629(a), which specifically states, "No person may dispatch or release an aircraft, continue to operate an aircraft en route, or land an aircraft when in the opinion of the pilot in command or aircraft dispatcher, icing conditions are expected or met that might adversely affect the safety of the flight," by failing to judiciously estimate the potential impact on the safety of our flight should we encounter the reported freezing precipitation that she knew was currently occurring just 10 miles from our destination airport. This lack of judgment, and the improper disavowal of responsibility and authority, was especially disconcerting at a time so close to the loss of Flight 4184 and in light of the new company directives and freshly stated FAA rules that specifically precluded us from operating our ATRs in *any* type of freezing rain or drizzle—no matter what the reported intensity.

Tim and I were restless and anxious. We had been out of the gate for more than a half hour, and we knew somebody needed to make a decision. During a short discussion, we evaluated our options and determined we would always have an exit from a dangerous icing encounter by turning back toward the warmer air that covered the southern half of Illinois. If, that is, we determined the danger early enough in the flight. That decision, we agreed, would be made at the slightest indication of ice, although we were not much nearer a clear determination to depart.

"What do you think?" I asked.

"Let's blast. We can always come back here if it gets crappy."

"Fine."

We got our altitude of 4000 feet approved to Chicago O'Hare, but weather diversions changed the flight, normally 40 minutes, into nearly an hour and a half. On the approach to Chicago O'Hare, we got the first of the ice on the flight. The entire flight had showed an air temperature of about 40 degrees at 4000 feet. On the glideslope for runway 9R, I checked it again at 2500 feet in moderate rain and saw a temperature of 29 degrees and asked the controller about the ground temperature, which still hovered around 40 degrees. We had found the inversion, and if we were going to get freezing rain, it was now. I was on the edge of my seat as Tim hand-flew the approach. After what seemed an eternity, the clouds parted, and we were in the clear with the runway about a mile ahead. We both exhaled a sigh of relief as I rechecked the temperature and saw it was back to around 40 degrees. We would be okay. We parked at gate G3 and were scheduled for a turn to Columbus, Ohio. As Tim shut down the engines, I began packing my bag.

"What's up?" he asked.

"I am not flying any more today," I said. "This is bullshit weather and I am not going to do it. Not after last month." I was inside and on the phone to scheduling before Tim and the flight attendants were off the aircraft.

"It's Steve Fredrick, and I want you to know I am done flying today."

"Weather related?"

"Yep." I looked up to see Tim and the flight attendants standing by the door. Motioning to them, I asked if they wanted out as well. They did. "And the rest of the crew."

"Okay, we have to put your names on a list."

"Whatever. See you. Anybody else?" I looked around to see several hands reaching for the receiver. About a dozen pilots who we personally talked to that morning refused to fly.

Nobody relished the decision or the ramifications. However, we had finally begun to take our fate into our hands. I sat in the terminal organizing my thoughts. The ramp outside the windows was crowded with ATRs. As I watched the cold November rain falling from the gray sky and rolling down the glass, I wondered what repercussions would arise from my refusing to fly the rest of my schedule. I felt melancholy as I watched the crowds pressing for answers at the G3 ticket counter. Apparently people would get on any airplane, any time. If these people knew or understood the passion and tone of conversations going on underneath the floor on which they were standing, would they be as quick to insist they had to get on that flight home? I don't think so. To the relief of the frazzled gate agents, some kind soul made the announcement, "American Eagle is announcing the delay of flights to Lansing, Kalamazoo, Cedar Rapids, South Bend, Rockford, Champaign. . . ." The list contained almost every one of our outstations. I wanted to hang around to see how things progressed, but I also wanted to get home. It had been a long month, and I was due back sooner than I wanted to contemplate. I noticed that the flight to Milwaukee was one of the few still scheduled to depart on time. It was being flown by one of the "heroes" from downstairs—one of the pilots who would prove there was nothing wrong with the aircraft. I knew I could get a seat on the airplane if I wanted it. I groaned and headed for a bus. As I passed the stairwell to the G3 loading area, the agent was taking tickets from passengers going to the Milwaukee aircraft. It was the only airplane on our side of the tarmac showing activity. The public address system again announced, "American Eagle regrets to announce the cancellations of its flights to Lansing, Kalamazoo, Cedar Rapids, South Bend,

Rockford, Champaign. . . . The reason for the cancellations is due to a lack of equipment." The ramp was full of ATRs.

Meetings occurred the following week, and more were conducted after the media attacked the airline. I was told by one of the captains in attendance that during a meeting Lance McDonald was paged by Robert Crandall, the chairman of AMR Corporation, who reportedly said, "Either get your pilots in line or I will shut down your operation." Surprisingly, Lance told the attendees of the conversation obviously meant only for his ears in an apparent attempt to intimidate compliance by the fear of the loss of their jobs. The tactic didn't work. The response from most of the senior pilots was relayed by one of my friends: "This place is not worth dying for, and if it shuts down because of this issue, so be it. I can't deal with the operational stress and your problems, too. If you guys think these airplanes are so safe, why don't you strap on an ATR and show us the way? Until you are willing to do that, keep your nose out of my cockpit decisions."

Shock waves were felt over the entire American Eagle system for months. Although damage control began almost immediately, it was too late. The story about the pilots who had been instructed to seek counseling made the newspapers, as did what the press dubbed "the pilot revolt."

American Eagle had another reported uncommanded right-roll event on Monday, November 28, when Flight 4052 approached Chicago O'Hare for a landing on runway 27L after less than 10 minutes of exposure to light-to-moderate icing. It was labeled a "turbulence-induced event." Bowing to the increasing upper-management pressure to keep the pilots in line and media scrutiny over the second-guessing of qualified pilots, Lance McDonald issued the following memorandum:

> *Due to the recent ATR problems, the following policy is effective immediately. . . . Pilots who do not wish to fly the ATR should contact their flight department manager as soon as possible so that appropriate arrangements can be made. Administrative procedures governing these reassignments will be forthcoming shortly.*

We saw the maneuver to placate our concerns as a grasping for straws that was rooted in a misunderstanding of the pilot character. Few, if any, crewmembers wanted to uproot their lives to go to another base to fly a different aircraft for the same money we were paid at Eagle; more importantly, it is generally not the nature of pilots to knowingly walk away from a problem that will inevitably be laid at the feet of other unsuspecting aviators. We would handle the situation in the cockpits. On November 30, Simmons Airlines again updated the release message pertaining to the ATR operating procedures.

8

Media involvement becomes necessary

Contacting the media regarding the crash of Flight 4184 was something I did not contemplate lightly. After being hired by Simmons, we learned early in ground school the airline's policy that all public communications were to come from the appropriate company spokesperson. It had been my dream to fly airliners all my life, and I was thrilled to be flying with American Eagle. Placing my career on the line was a minor consideration, however, because I had larger issues guiding my decision. I had a duty to my employer, but my employer had a duty to its employees and the flying public. In my opinion, the company had been derelict in that duty for years. The ATR was well-known in airline circles. We talked about its flaws and that it was only a matter of time before the worst happened. Yet the passengers kept coming, the fleet kept growing, and we kept flying the airplane in all weather. After the crash, we saw the same pattern of minimizing problems and past incidents that we had seen before the crash, and I made a promise to my fallen coworkers.

During the first days after the accident, I monitored press coverage, searching for any word pertaining to the accident or investigation. I watched to see who was writing accurate reports and who was "aeronautically illiterate." Few journalists take the time necessary to learn the intricacies of aviation terminology to avoid the obvious errors seen by those in the industry, but Gary Washburn of the *Chicago*

Tribune seemed the most accurate. Unlike reporters who wrote about the "flaps turning the airplane," Gary had few, if any, erroneous statements. He would be the one to contact, if and when the time came.

I called the *Tribune* to introduce myself to Gary in early November and told him I would like to be able to contact him if I saw things going awry. During that initial conversation, I gave him a cursory overview of the history of the ATR and told him how to get in touch with me, but I was not active in the dissemination of detailed information until later in the month. I was feeling him out as much as he was doing the same with me.

With the decision made, I was determined not to sensationalize the story nor indulge in righteous fulmination. What needed to be publicized was a clearly defined, well-organized statement detailing the checkered history of the aircraft and the recurring patchwork fixes established by the manufacturer and sanctioned by the FAA. The crash was painful enough, especially to the victims' families, and I was determined not to add to that pain by making reckless statements.

This particular November passed with a surreal slowness. Each day was packed with controversy, modifications to the operating procedures, wisps of inside information regarding the accident investigation, management and pilot flareups, and critical stress. The crew lounge was ready to explode, and cockpits were packed with tension, especially those that encountered ice during the flight.

Pilots in the crew lounge began to congregate into small clusters of hushed discussion; everyone was concerned with being overheard by someone who might report the conversation to management. Seeing groups of uniformed aviators huddling in a corner engaged in quiet discussion, with one occasionally turning his head to see whether anyone was watching, reminded me of a group of prisoners plotting a jailbreak while watching for the guards. I have never witnessed anything like it in my life, outside of movies, and I joined these cloistered conversations whenever I could.

All the pilots I spoke with realized we would either see the FAA moving for a final fix once and for all, or we would experience another sugar-coating of the problem. We hoped for a fix, but we expected and began to see more of the same type of lip service and return to operations as usual. We understood that public attention would last only for a limited time. It would take a quick, strong action to change the path we were traveling, and the pilots' union seemed more concerned with jobs than the performance of the ATRs in ice. You could feel the tension when you walked into the crew lounge, and many of us were afraid the stress would cause somebody to do something stupid if nothing changed soon. By mid-November, the time had come to take drastic action; the situation found me before I could act.

The day after I and other pilots had refused to fly, I was in my office in Elkhart Lake, Wisconsin, and the sun was shining through broken clouds; it was a far cry from the day before. The call came in the late morning or shortly after lunch. I was surprised to hear from Gary Washburn; I had placed the events of the past frantic Thanksgiving holiday weekend out of mind.

"Steve, I understand American Eagle had a number of flight cancellations over the weekend, and I wondered if you were flying and could add anything to what I already know." I could hear the clicking of the keyboard as he typed our discussion notes directly into his computer terminal at the *Chicago Tribune.*

"I don't know. I was flying, and the weather turned really lousy on Sunday, with freezing rain reports. I know I refused to fly when I got back from Champaign, Illinois."

"We've got reports that there were about a dozen cancellations, and American Eagle is stating they delayed the flights because pilots wanted more weather information and by the time they got the information it was too late to send the flight so they canceled it. AMR spokespeople have confirmed the cancellations but insist no pilots refused to fly."

"Well, that's just bullshit," I exploded. "I was there, I was scheduled to fly and I was talking with at least a dozen other pilots who said they had already told crew scheduling they weren't going anywhere. For them to say nobody refused to fly is a blatant lie."

The story appeared on the front page of the *Chicago Tribune* on November 30, 1994. Titled "American Eagle pilots refuse to fly—several ground selves in cold, wet weather" the article included a number of quotes from other anonymous Eagle pilots, many of whom I recognized from their statements. I was glad Gary had other sources. Two other pilots expressed their concern for the company and their overriding concern for safety. One said, "I am an employee of this company, and I love my job. I have already made up my mind not to fly if faced with unsafe weather conditions. The last thing I want to see is the company fail as a result of something like this. This could cost me my job ultimately. But certain things have to be put in order."

The second pilot said, "We're in a bad spot here, we really are. We understand the company has to stay in business, but until they resolve the problem with this airplane, we need to be a little more picky about the weather we fly in. We have public safety in mind and also our own."

I saw my own words in the article describing the weather of the past Sunday: "The weather conditions were very similar to the day the accident occurred. I spoke with at least a dozen pilots who had refused to fly. Conditions were favorable for freezing rain to be in the area, and the first report of freezing rain came from Rockford. Based on that, a lot of the pilots refused to fly, feeling that was close enough to Chicago that they weren't going to take any chances."

Another coworker spoke of pressures applied by management to intimidate the pilots when he stated, "When I was there, management was on the phone and wanted to know, one by one, who was refusing to go and who wasn't."

Gary patched together a conclusion from pilot statements and the spin control by the airline in the next paragraph of his story. He wrote, "The severity of weather conditions, in part, can be a matter of perception. One pilot may believe conditions are potentially hazardous, while another may conclude they are manageable and present minimal risk."

That passage reminded me of two stories that related to that conclusion. Smart pilots become good pilots. For other pilots, their ego, if it does not kill them, often places them in dangerous positions or causes future problems. The first story involved a pilot who was known to be a "bad stick" and only made captain in the Shorts 360 at our airline because of seniority with the company. One day, a line of convective activity extended from Canada down to Arkansas, and it lay across the flight path from Chicago to Green Bay. The weather radar in the operations area showed the thunderstorms had grown into a solid line of heavy activity. The movement showed they would pass through Chicago in a couple of hours, and most north- and westbound crews canceled their flights. The captain who was scheduled to fly to Green Bay decided he did not want to tangle with the weather. Overhearing that, this other pilot called crew scheduling and insisted on taking the flight to Green Bay; she got them to change the cancellation moments before it was official.

My future wife was the scheduled flight attendant on that trip, and we were standing in the crew lounge together. I looked at the map and said to the pilot, "Are you sure you want to do this?" The pilot looked at me like I was an idiot. They departed, the only flight going out during that particular block of flights. An hour and a half later, which was longer than the scheduled flight, the aircraft landed back in Chicago with at least one frightened flight attendant. Linde told me they had "the crap kicked out of them as they kept trying to go through the weather." People got sick, people got scared; and after all that, they were back in Chicago anyway.

The second story involved the happenings during November 1994. An entire ATR crew decided the weather conditions were not conducive to the safe operation of the flight because of the ice encountered inbound to Chicago. They called crew scheduling to state their refusal to fly, and to their surprise, it was accepted. Unknown to the crew until later, the company tapped three new crewmembers sitting in ready reserve at the airport and assigned them to the trip without advising them as to why the first crew was not doing the flight. Because they had been sitting in the crew lounge all day and had not experienced the weather conditions, this reserve crew went on its way. Although nothing untoward happened that we know about, crew, aircraft, and passengers were placed in jeopardy.

American Eagle spokesman Marty Heires, who now works for Greyhound Bus Lines, denied management was putting undue pressure on the pilots to fly. "I will categorically deny we have forced any crews to take flights in unsafe weather conditions."

The game of semantics had begun. Word wars waged for months in the media, and every statement was countered, rebutted, and twisted. American Eagle, ATR, and even the FAA had their own professional, highly skilled public relations staffs, but the pilots had only the truth on our side. We were waging a battle against well-funded, well-connected, and powerful organizations, but we had the ear of respected, reasonable, and honorable members of the media.

On the same day the first controversial article appeared in the *Chicago Tribune*, the phone rang at my office in Elkhart Lake. Again it was Gary Washburn; this time he had an interesting request. The article had intrigued the producers of the ABC television show *Good Morning, America.* They were looking for a pilot to talk to them on the next day's show. I agreed to talk with them and a short time later, I received a call from Rebecca Levy. Rebecca told me the show wanted to do a live interview of an American Eagle pilot and asked if I could make it to New York for the next morning's show. When I responded in the affirmative, I was told to expect a call from a writer for a quick interview to formulate possible questions and a fax providing the information for the travel and accommodations. Before she hung up, Rebecca asked about my feelings for security.

"We understand you are still anonymous," she started. "We can disguise your identity with a voice mask, special makeup, or an appearance in silhouette. Or we can do all three."

"I guess I haven't thought about it much," I responded, remembering Gary's advice that American Eagle would do everything possible to discover my identity after the appearance. "I have seen the disguises used before, and I find them to be a problem with the person's credibility for me." Finally, we settled on a silhouette appearance.

When the writer contacted me a short time later, one of his first questions was "What should I refer to you as, first officer, copilot. . . ."

"Just call me Steve," I answered, thinking he meant for the purposes of the conversation.

"Okay, Steve it is," he replied and then continued with the short interview.

Two hours later, I was on my way to General Mitchell International Airport in Milwaukee, just as I had done hundreds of times before to catch a flight to Chicago to go fly for American Eagle. Only this time I felt different. My stomach churned as I walked up to the Continental Airlines ticket counter, but I was thankful the show did not book me on American. It was easy to remind myself that I was doing the right thing; it was another matter to fully accept where this could take me. Continental Airlines departs from the C concourse in Milwaukee, beyond American's gates, and the security people recognized me before I entered the area. American Eagle was boarding a flight as I walked by, and the agents watched me pass, most likely wondering where I was going, but they didn't ask. I anticipated a more uncomfortable passage on my return the next day. My thoughts were a jumbled mess as I waited for the flight to board, and I do not remember much of the ride to Newark.

I caught a cab downtown and checked into the hotel. The check-in register contained the name of Rebecca Levy, and the room was identified as the *Good Morning, America* Permanent Suite. I wondered how many celebrities had stayed in the room. Butterflies flitted in my stomach.

A car was to pick me up in the morning and take me to the studio. I nervously paced in the lobby of the hotel as drivers came and left with their charges, but nobody asked for me. It was now becoming close to the time I was to be at the studio, and my concern grew. I checked with the doorman to see whether any of the half dozen cars parked on the street were for me. I frantically called the number at the studio that Rebecca had provided me in case any problems arose. Within minutes, a driver appeared in the lobby. He had been waiting for Rebecca Levy. It had been part of the security. It almost cost what was left of my composure.

The driver stopped on a deserted street before a nondescript door in the back of a nondescript building. I exited the limousine and was immediately greeted by a frantic woman who whisked me into the studio "green room." I was glad I was going to be in silhouette so the audience probably could not see who I was. "No makeup needed," the woman said, and I went into the hall to meet Charlie Gibson, one of the cohosts. It was 10 minutes to show time, and the activity was low-key but hectic. Charlie took the time to talk with me about the interview, telling me to relax and enjoy myself, and showed me the beige leather chair I would be seated in. A man with headphones told me to keep my answers short and on topic and to respond to Charlie. I guessed he was important, so I nodded in agreement. Soon I was left alone as the host moved to his position to begin the program, and I heard the reference to my appearance. I wondered whether my family was awake for the show.

When the show began, I was informed I would be on just after the news and weather, so the crew got me seated and set up with a microphone. A large-screen television showed my silhouetted form through the camera placed 10 feet in front of me. I looked at the screen and noticed the characteristic shape of my head and a large, distinctive shiny spot high on my forehead. "Oh boy," I thought. "This is going to be real tough to guess who I am." It was too late to change anything, so I just sat there and thought about what I would say. On another monitor, I could see the newscast that included a report on the crash and the pilot action of the previous Sunday.

Charlie began the introduction of the piece, and I straightened in my seat. The last quick drink of water I had taken had gone down the wrong pipe, and I was fighting a coughing spell. My palms were sweating, and I knew instinctively it was too late to clear my throat or test my voice. It was anybody's guess if I would open my mouth with the sound of a squeaky-frog voice of a 14-year-old.

"As Lisa Stark reported in the news, a group of American Eagle pilots are grounding themselves, refusing to fly their planes in icy conditions similar to those that caused an American Eagle commuter flight to crash in Indiana a month ago. Joining me this morning is one of those pilots. He doesn't want his identity revealed, so we'll just call him Steve, and you will see him only in silhouette this morning. American Eagle declined our invitation to appear this morning. . . ."

"Just call him STEVE!" my mind screamed. What the heck is that all about? I knew I was busted by the company as soon as the discussion started. "Oh, well," I thought, "I knew this could happen."

The next five minutes were a blur of conversation and attempts to make clear points. The appearance would not win any awards, but the controversy was laid squarely into the public forum. Almost as quickly as it began, it was over, and I found myself chatting with Charlie in the hall by the green room again as commercials ran. In another moment he was gone, and I was shuffled out the door into the waiting limousine that would take me to the airport in Newark. Laying on the back seat was a copy of the *New York Times*, and I found an article on the Eagle situation. I felt good about what I had done, but I had a feeling of foreboding that became stronger as we approached the airport. A Continental Airlines ATR-42 flew directly over the car and landed on the runway across the road. I felt as if a vulture had landed on my shoulder. To add insult to injury, Continental canceled the flight back to Milwaukee, but I gained a reprieve when they routed me on Midwest Express an hour later. Maybe things were looking up.

When I got home, a message was waiting for me to contact Erin Hayes, the Chicago-based ABC network correspondent. My name and phone number were getting out, although I still hoped for anonymity. I agreed to do an on-camera interview with Erin on Friday, December 2, as long as they still did not reveal my identity. I was beginning to learn about the credibility of the media and the meaning of some of their terminology. I learned early on, due to Gary Washburn's candor, that anything said to a reporter before receiving an agreement of the conversation being "off the record" is fair game for that reporter. As a matter of courtesy, Gary gave me an understanding of the rules when dealing with journalists. I knew my security would be protected by Gary under his code. I later found that Erin Hayes also held her code of ethics in high regard.

News comes from people who talk to reporters. I had never spoken to Erin before meeting her at the Pfister Hotel in downtown Milwaukee on that Friday, yet she had more inside information than I had, or so it seemed. When I walked into the hotel room with my wife, Linde, and ABC producer Jim Hill, Erin was on the telephone with another American Eagle pilot source, rapidly scribbling notes on another uncommanded roll event that occurred the previous Monday as the ATR approached Chicago O'Hare. We were interrupted twice more during the course of the two-hour interview as Erin returned calls to her pager from other Simmons ATR pilots. The reporter did not tell me who the other aviators were, and I would not think of asking—their security was as important to them as mine was to me—but I had a good idea as to their identities. As we wrapped up the interview session and headed for home, I felt like I was doing something to benefit the situation. Then Linde reminded me I was scheduled to fly the next day.

On Saturday, December 3, 1994, I had a late afternoon sign-in time for a four-day trip. I did not know who the captain was. I was determined to go with the flow and see what would come from my appearance on national television three days earlier. Walking past the cockpit of the ATR-72 that would take me from Milwaukee to Chicago O'Hare, I recognized the crew and waved a hello. I received a smile and wave back.

I cannot describe the feelings I had on arriving at Chicago O'Hare. I did not know whether I would be met at the flight by management personnel, whether I was recognized by passengers or my fellow employees, or how I would react to conversations or questioning. I soon found out.

When I walked into the operations area of the crew lounge, the first person I saw was Captain Larry Blacksmith, who smiled and greeted me with a hearty, "Hey, Stevie!" The look on his face and his smile told me the whole story. It was no secret. Everybody knew, without a doubt, that I had been the American Eagle pilot on *Good Morning, America*. The television had been on in the crew lounge the morning of my appearance, and reportedly the reaction included both cheers and jeers.

Then it came. The crew coordinator summoned me to the telephone to take a call from crew scheduling. My trip had been canceled, and I was assigned to fly roundtrip to Rockford, Illinois, and then an overnight in Evansville, Indiana. I would be with Captain Grubb, whoever he was, and it would be in an ATR-42, not my scheduled ATR-72. I hung up. The area had cleared out except for a couple of sleeping flight attendants who were sitting ready reserve. I felt alone and called Linde to tell her, "They know."

I walked around the terminal and wandered into the newsstand where I had spotted the now infamous leaflets the week before. The Sunday edition of the *Chicago Tribune* was already on the shelf, and I was stunned to see a large photo of an American Eagle ATR-72 and a front page, above-the-fold headline reading, "Pilots, American Eagle clash over safety of ATR turboprops." Gary Washburn, writing with Theo Francis, had blacked another eye of the ATR. Another voice, the 110,000 members of the International Airline Passengers Association, had joined the debate over the safety of the airliner and advised "stay off ATRs whenever possible." Travel agents were reporting 85 percent of customers were asking questions about the types of planes on which they would be booked. The FAA announced that their special certification review of the ATR would complete testing by February 1, 1995. I purchased the paper and placed it in my computer case.

In the crew lounge, I met Captain Grubb. He wanted to talk to me privately. We went into the bag room and closed the door.

"Are you the guy who was on television the other day?" he snapped.

"Listen, I don't know you from Adam, and I think that question is a little out of line," I responded indignantly.

"I don't like people messing with my job. If you don't like it here, then why don't you just leave?"

I held my composure although I really wanted to give this guy a good lecture. "How much time do you have in this airplane in the ice?"

"What does that matter? I just got out of training. I have been flying in Dallas, and I have heard the stories."

"Well, I have nearly 4000 hours in the ATR, flying in the north. I knew the crew who were lost in 4184, I was hired with the first officer. Don't try and get into my mind and figure you understand my motivations. I want this airplane fixed. I have seen the problems swept under the carpet too many times before." My finger was pointed at his face, and the lights of the room were reflected in his eyes just over a foot away from mine. "Until you have the time and experience in this particular airplane that I do and that the other Chicago pilots have, I suggest you watch, learn, and don't rush to judgment; the whole time pray that you don't have an icing event yourself."

"So were you on television or not?" he pressed.

"What difference does it make?" I answered. "I don't owe you anything, least of all an explanation for what I may have or may not have done."

"If I could prove it was you, I would refuse to fly with you."

"Make the call." I left the room.

We flew the turn to Rockford together, and I tried to break the cockpit tension on the way back by talking about other subjects. As he left the cockpit on our return to Chicago O'Hare, I did not realize I would never see Captain Grubb again.

On arrival back at Chicago O'Hare, a nagging apprehension developed as I walked across the ramp through the chilled night air. I knew AMR would not let me get away with publicly humiliating the company. It was after 7:00 P.M. on a Saturday night; finding one management person at the airport at that time would have sounded an alarm. Tonight there were two, and they were accompanied by a man I did not recognize. I was stopped by Ed Simon, the chief pilot.

"Steve Fredrick?" he inquired.

"Yeah," I answered.

"We want to do a flight kit check as allowed under the ALPA contract. Would you please come back to your aircraft with us and collect your belongings?"

"Okay."

We returned to the aircraft, and I walked into the cockpit to gather my flight kit, which contained my checklists, manuals, and charts. I placed the black leather case near the open cargo door, and the man I didn't recognize went into the cockpit to look around. I assume he was looking to see if I had left anything behind, which I had not.

"Bring your luggage, too," Ed Harvey said. "We want to check everything."

"No problem."

We walked together into the lower level of the concourse, an area I had not seen before, and took an elevator to the main level. We went into the chief pilot's office, located between gates G3 and G5.

"Please open your flight kit and take everything out," Ed Simon directed. "Just put the stuff on the desk."

I complied and stepped back as they checked the revision status of my navigation charts and two required company manuals. My checklists were verified, as was my flashlight. Everything was in order, which I knew before they began. I replaced the items back into my bag, which Simon had overturned to make sure there were no other items hidden in it.

"Okay?" I asked.

"Now we want to look through your personal luggage."

"Why?" No reason existed for me to allow the search except for the fact that I was faced with three hostile management types who knew I had just talked out of school about the company.

"Just open the bags," Ed Simon stated.

"What are you looking for?" I asked, knowing they would not tell me.

"Contraband."

"Contraband?"

"Yes, contraband. Now I am directing you to open your bags."

"What kind of contraband?"

"Papers."

"Papers are contraband?"

"Yes, they can be."

"I don't think I am going to allow you to look through my personal bags," I answered. "It is an invasion of my privacy. Besides, these bags have been security screened, and you have not given me a good enough reason to allow a search."

I could see they were getting irritated. "I'm ordering you to open the bags," Ed Simon snarled.

"No," I flatly stated.

"We will get the Chicago Police, and they will arrest you and then we will open the bags."

"Go ahead. Get the police, get security, get anybody in authority in here, and I will open the bags and allow a proper search. But you guys are not rummaging through my personal belongings without a better reason than 'looking for papers'." I stood my ground. I believed I was right and suspected they were bluffing because they started looking at each other, waiting for someone to make a move.

We were getting nowhere. I maintained my position and so did they. They gave me one last direct instruction to open the bags, and I refused. They informed me I was suspended without pay and ordered me to surrender all company identification and keys. They typed out a short notice that was signed by Chief Pilot Simon and initialed by Ed Harvey. I collected my bags, said good night, and left the room with my dignity intact.

The next morning, I returned to the airport dressed in my civilian clothes and purchased a ticket on the next bus to Milwaukee. For the second Sunday in a row, I did not return from Chicago on an airplane.

I knew there would be further repercussions. I also knew I would continue the fight to make the ATR problems known to the public, from the outside if necessary. The bus rolled out of the airport, and I watched the sights fondly, sadly realizing I would never see them again from the perspective of an airline pilot.

Pilots and flight attendants were not just nervous, we were scared. The consensus was that something needed to be done, but no leaders stepped forward because of the fear of company reprisals. The public perception of airline pilots and flight attendants is that they do not work much or often and that they make a lot of money. Maybe at the major airlines that is the case, at least with regard to money, but on the regional level, crews are blue-collar workers. Great media outcries came out of the American Airlines flight attendants' strike of November 1993 regarding the fact that starting pay for these hard-working professionals was only $13,000 annually. At American Eagle, flight attendants started out with a base salary of less than $10,000, and some even qualify for food stamps and other government assistance. When the airline hears of a pilot or flight attendant who needs to supplement his or her income with government assistance, they bring the employee into the supervisor's office and advise him or her that receiving assistance is counter to the public image that American Airlines wants to project, and the employee is pressured to drop the program and get a second job to make ends meet. Our flight attendant's maximum salary is about $17,000 per year without overtime pay.

New-hire pilots at Simmons earn about $12,500 per year base salary, and the top pay for first officers is less than $25,000 in wages and overnight reimbursements each year. Even captains, who across the alley at American earn a six-figure income, range from $25,000 to around $50,000 each year at the Eagle carriers. With overtime, some earn more. American Eagle employees are definitely middle- and lower middle-class wage earners who need their jobs. We love aviation and are willing to make sacrifices to work in the profession, but we have families to support and could not afford to challenge AMR Corporation.

During the first week of December 1994, public opinion was turning against the ATR airliners as a result of intense media attention. Travel agents reported drops in bookings on ATR flights, and many customers were still asking questions about the types of aircraft they would be traveling on. Public knowledge of airplanes and the hazards of ice was expanding, and I wondered how long the parties who were the focus of the frenzy would take it before striking back. I did not have to wait long.

I had become a sought-after spokesman for the story on the ATR. No one can be prepared for the onslaught of media attention when thrust into the spotlight. I had no desire to become publicly known and first made contact with the print media as a mostly anonymous source. I believed intervention was necessary because I saw the accident and the aircraft problems being ignored. This time I had lost friends, and the rules had changed. No longer could loyalty to the company, the desire to fix things in-house without public attention, the potential for public overreaction, and the financial harm to the airline that occurs in such situations mold my thinking. My duty had changed—I had no other choice.

Simmons Airlines enlisted several pilots to give interviews in a "counter-Steve" campaign to build back the public confidence and proclaim the ATR aircraft safe. Two of the group, Captains Larry Blacksmith and Chuck Tolleson, could honestly state they had not had any serious ice-related incidents in the ATR, and Chuck proclaimed the ATR totally safe and without need of further FAA action.

On December 9, the FAA grounded the ATR in known or forecast icing conditions, effectively shutting down the scheduled ATR operations of airlines of the northern United States.

The crew lounge hushed as the news came over the television. Chuck Tolleson had been telling other pilots about his interview when the story broke. A fellow pilot told me he watched Tolleson more than the news and described the scene to me later. As people turned and watched the television, Tolleson backed out of the group and sat down, holding his head in his hands. When the reporter had finished his story, a voice called from the far corner of the room, "Hey, Chuck, how do you feel now?" The laughter at Chuck's expense was short as everyone wondered what the grounding meant for them. As the flight crewmembers began leaving Chicago O'Hare, their aircraft useless on the ramp, reporters surrounded the dejected pilots and wanted their reaction. I am told by one friend the oft-repeated question was "Do you know Steve Fredrick?"

On the day the aircraft was grounded, I arrived back in Milwaukee from New York and another appearance on a talk show. My office caught me before I left the airport and gave me the news, which was not yet official, but media contacts had reported that other sources inside the FAA indicated the flight limitation announcement was going to come at a press conference later in the afternoon. I was asked to provide comments to all three major networks, and I spent the rest of the afternoon rushing between stations in the Milwaukee area. My goal was to get the word out via the most avenues as possible.

When the final announcement came, I was doing a live feed to the network news at CBS and was asked if I felt vindicated by the announcement. That question had been asked numerous times that day, and my only response was, in essence, "I'm elated that they finally decided to protect the flying public and do what they should have done after the crash."

Al Gagliano, the father of First Officer Jeff Gagliano, also was outspoken during this period and summed up the entire story in a simple statement that reminded us all what this was about: "I'm angry that the FAA and the company that built the planes let this go on. It took the deaths of 68 people, not just the death of my son, to bring this to a head." The interview was given on what would have been Jeff's 31st birthday, three days after the grounding of the ATR.

The evening after the grounding, I received a telephone call at my home. CNN reported the loss of another American Eagle aircraft, this one on approach to the Raleigh-Durham Airport in North Carolina. It was a horrible coincidence, and I made statements in several interviews attempting to bolster the public confidence in the American Eagle system. I saw no connection between the two accidents, and a rush to judgment by some in the media regarding the entire regional airline system had to be derailed. I did what I could to that end, regardless of my employment situation with American Eagle.

I hold no animosity toward my coworkers who failed to come out publicly in my support. I understand their position financially and am grateful for their continued, behind-the-scenes support. Nor do I resent anyone who spoke out about the positive attributes of the ATR aircraft to undermine my efforts. Throughout my associations with the media, I tried to maintain the focus of my statements on the ATR and not delve into peripheral sensational issues.

As elated as I was on December 9, when the aircraft were taken out of service, I was equally deflated 33 days later when the ban was lifted. I believe the increased public knowledge about the aircraft led the FAA to ground the aircraft, although the agency later discounted public scrutiny as a catalyst for action. Meeting notes from the FAA proved the proactive stance taken by the agency with regard to the public image of the ATRs. These same notes also showed the statements of the airlines of going out of business if the airplanes were not back in the air by mid-January, and by January the ATR returned to the air.

The story was beginning to take a toll on my family, and Linde wondered when it would end. Then a call came from Marian MacNare of the Canadian Broadcasting Company in Montreal seeking my help with a documentary that producer Diane Ngui-yen was assembling on the life and untimely death of Bill Readings, a 34-year-old professor at the University of Montreal. I was asked to travel to the city to give an interview for the piece and I agreed. Simmons had changed my status to paid administrative leave, so I had my travel benefits and company identification. I hitched a jumpseat ride on Northwest Airlines. It was a long day of traveling, and the interview went until nearly dark. Diane took me to a hotel in downtown Montreal, and I dragged myself to my room. I was exhausted, mentally frazzled, lonely, and my head ached, so I decided to take a hot bath and soak away the aches and pains.

Sitting in the hot bath, I remembered the story Diane had told me on the way to the hotel. Bill had been visiting his wife, also named Diane, in Bloomington, Indiana, the weekend before the accident and was returning home on Flight 4184. He was due at a Halloween party, and I pictured the guests dressed in costumes waiting at a restaurant in Montreal for Bill to arrive. They received a call from Bill's frantic wife in Indiana, who was hoping for a mistake. "Is Bill there? I think there has been an accident with Bill's plane."

As I played the tale again in my mind, I realized I had to tell the human side of this story as well. I had been concentrating on the technical side of the tragedy, but this scenario had played out hundreds of times that day as friends and relatives gathered to hear the fate of those on American Eagle Flight 4184. Their stories needed to be told.

9

Unheeded
warnings of
disaster

After deregulation of the airlines occurred in the United States in 1978, the industry changed rapidly. New carriers began operations and shut down with alarming frequency. The established major airlines restructured or eliminated unprofitable routes to survive while they fought to expand service to new markets previously blocked to them by government regulation.

To improve public perception of the small regional carriers, United, American, Northwest, and USAir, among others, began lending their company colors, adapted logos, services and equipment, and computer reservation systems to selected regional partners. Code sharing, which allows passengers to continue travel from a major airline to a regional airline on one ticket, began and continues today.

Although many small cities generated a moderate volume of air travelers, it was not enough to justify expensive jet service. The recycled business aircraft and 19-passenger, first-generation commuter aircraft used by the fledgling commuter airlines were limited in capacity, safety margins, and passenger comfort. If the market was to flourish, a new aircraft design was needed.

In 1981, Avions de Transport Regional, known as ATR, was formed by the French company Aerospatiale and the Italian firm Alenia. The consortium, based in Toulouse, France, was founded to develop and produce a new, twin-engine, turboprop airliner. In February 1982, ATR applied with the FAA for an import-type certificate for its ATR-42 airplane. The new company moved forward with the ATR-42 project rapidly, and on August 16, 1984, the first of two developmental aircraft flew over southern France.

A new era of airline travel began with the introduction of the ATR to the marketplace. The ATR-42 filled the gap between the small commuter aircraft, which still served the smallest markets, and the expensive-to-operate turbojet aircraft. The ATR aircraft is cost-efficient to operate, passenger-friendly, and uses state-of-the-art cockpit instrumentation and design. The ATR-42 offered passengers many of the amenities they had become accustomed to in jet service, including an aisle in which they could stand up, overhead storage bins, and a quiet, roomy cabin. The ATR helped change the image of commuter airlines and transformed them into more respected regional airlines.

All ATR-42 models have a wing span of 80 feet, a length from nose to tail cone of 74 feet, and a height to the top of the tail of 25 feet. The aircraft is powered by two efficient Pratt & Whitney PW120 turboprop engines, each of which produces 1800 horsepower under normal conditions, with a total reserve takeoff power of 2000 horsepower for emergencies. These new engines transmitted power through a reduction gearbox to four-bladed, 14-foot-diameter propellers that turn at a maximum of 1200 revolutions per minute. Normally, propeller speeds are rated in percentage of maximum RPM, with 100 percent used for takeoff and final approach to allow for rapid spooling of the engines in case the control tower orders a go-around; 86 percent used for climbing, high-performance cruising, and all icing operations; and 77 percent for a slightly less efficient but quieter normal cruise flight. The blades of the propeller are made from a composite material with a limited life, and each of the eight blades per aircraft costs nearly $75,000 to replace.

An ATR-42 equipped for passenger service weighs between 23,000 and 24,000 pounds when empty and has a maximum takeoff weight of 36,825 pounds. Unlike many turboprop aircraft, the ATR-42 can operate out of large airports with 46 passengers (the maximum number allowed in the Simmons Airlines-American Eagle configuration), more than a ton of baggage and cargo, and enough fuel for a 500-mile flight, with reserves, even on the hottest days when lift on takeoff is poor.

As is the case with all new airplanes, the ATR-42 suffered mechanical problems when placed in normal service. For the first year at Simmons, ATR pilots joked that they spent more time ferrying the aircraft to the maintenance base at Marquette, Michigan, than they did transporting paying passengers. One ATR first officer, in my class as an upgrade to ATR captain when I was a new employee, told me he wanted to be assigned to the aircraft because it gave him ample time off due to flight cancellations.

In mid-1985, nearly a year before the ATR-42 went into service with the first two U.S. carriers, the ATR consortium announced the decision to develop the ATR-72, a stretched version of the ATR-42. Several improved versions of the two models of the ATR aircraft exist, but the 46-to-50-passenger ATR-42 and the newer ATR-72, which seats up to 70 passengers, remain basically the same as when they were introduced.

The ATR consortium also investigated the feasibility of another stretch of the ATR aircraft or a possible fanjet-powered model to serve longer routes, either of which might have been christened the ATR-82. After the crash of American Eagle Flight 4184, however, the rumor circulating in the crew lounge at American Eagle indicated that ATR had temporarily suspended development of future versions of the aircraft.

Finnair, the national airline of Finland, was the launch customer for the all-new design and received its first ATR-42 on December 2, 1985. Command Airways, subsequently purchased by AMR Eagle and merged with Nashville Eagle to form Flagship Airlines, was the first U.S. operator and received the aircraft shortly after the delivery to Finnair.

Simmons Airlines, founded by Larry Simmons in Marquette, Michigan, in 1978, started out flying piston-powered, twin-engine corporate aircraft, mainly serving the just-in-time parts philosophy of the automotive industry. It later became a connecting carrier with American Airlines in Chicago and Northwest Airlines in Detroit and Minneapolis, providing continuing service to smaller communities. Simmons prospered with the new opportunities offered by this arrangement and grew under the guidance of its later owners, the Murray brothers. The airline was sold to AMR Corporation, parent company of American Airlines, in 1988 and has operated as a wholly owned subsidiary of American ever since. AMR has invested heavily in its regional partners and takes pride in the seamless service it can offer.

The second scheduled U.S. operator of the ATR-42, Simmons received its first aircraft in mid-1986. The company became the largest operator of ATR aircraft in the world, with more than 70 of the turboprop aircraft. FAA records indicate the aircraft type was added to the Simmons operations certificate on June 15, 1986, just nine days after the first reported icing-related control incident occurred in an ATR-42.

The ATR-72 was created by adding 15 feet of cabin to the ATR-42 fuselage and stretching the wingspan to 89 feet. The design process yielded a new model that is similar to its predecessor. The cockpits of the ATR-42 and ATR-72 have nearly identical layouts; the larger model has a few additional switches and buttons for its modified or new systems.

The ATR-72 aircraft are also powered by Pratt & Whitney turboprop engines, although a different model than that used on ATR-42s.

They produced 2400 horsepower on the earlier models and 2700 on the later versions. The newest model, the ATR-72-212, which was the model that operated as Flight 4184, uses a curved, scimitar propeller blade that reduces vibration and cabin noise and converts engine power to thrust more efficiently. At American Eagle, an upgrade was scheduled to an even newer model of propeller that used a six-blade configuration; however, the design ran into snags during testing. A design flaw in the hub of the propeller resulted in the loss of propeller blades. An aircraft would take off with 12 propeller blades but return with 11, distressing the test pilots and necessitating a trip by ATR technicians into the French countryside to locate and recover the blade.

The ATR-72 uses a sophisticated multifunction computer that monitors and automatically controls many of the aircraft operations. The computer removes some of the workload from the pilots by seamlessly switching or shedding electrical loads in case a problem is indicated by a sensor, requiring the crew only to verify that corrective action has occurred when reciting the checklist. To the dismay of many ATR pilots, however, the aircraft designers decided to use this computer to prevent pilots from intentionally or inadvertently exceeding some aircraft limitations and thereby eliminated some critical emergency options. Complaints to the airline and ATR yielded no positive results, so pilots learned to live with the computer control.

One of the most disconcerting of these designer-imposed preventive measures prevents pilots from moving the flaps from the retracted position to the 15-degree setting at speeds above 180 knots even though the maximum design speed for operating with the flaps at 15 degrees is an indicated airspeed of 185 knots. Unfortunately, no override of this system exists if an emergency should occur. Thus, ground schools teaching the ATR aircraft were often places of debate and discontent regarding the French design decisions. Even instructors would mock the French, who were noted for saying "Is no problem, is au-to-ma-tic," which would be mimicked in the best-possible accent.

Pilots like to use technology, but they also like to switch technology off if things go wrong and the pilot needs to fly the airplane instead of being a passenger. Few pilots envisioned the scenario that occurred over Roselawn, but if the pilots had been able to turn the restriction on flap settings off, the crew of Flight 4184 might have been able to regain control of the airplane. Although the crew would have exceeded a recommended operating restriction, they might have recovered after the upset occurred and the airspeed began increasing rapidly. After the crash, FAA mandate removed this restriction.

All models of the ATR aircraft are manufactured by both partner companies, with the Italians constructing the fuselage and tail section, which are shipped to Toulouse, France, for mating with the French-made wings, final assembly, and painting. Delivery of the finished product to the purchasing airline occurs at ATR's headquarters near Toulouse.

ATR enjoys some perks not available to corporations in the United States. Those special privileges occurred, in part, because ATR has two powerful partners, the French and Italian governments, which subsidize the company and are part owners. One of the defenses ATR is using in the liability lawsuits filed against the company because of the crash is immunity from prosecution as an entity of a foreign government. According to information available, 50 percent of the individual companies of the consortium are owned by their respective governments. Therefore, France and Italy each own 25 percent of ATR, with the remaining 50 percent of the company owned by yet-unnamed parties. ATR Marketing and ATR Support, corporations that operate out of offices in Chantilly, Virginia, near Washington, D.C., exist mostly on paper and have few assets of value to the families of victims.

Unlike airliners built in the United States, the ATR models, as well as aircraft of some other foreign builders, are not originally certified by the FAA. For example, a Boeing jetliner undergoes rigorous scrutiny by the independent government agency during almost every phase of its development, construction, and testing. The U.S. government spends thousands of personnel hours overseeing the new aircraft and determines that it meets all of the safety requirements before it is issued a type certificate. The type certificate allows the manufacturer to build and sell the aircraft within the United States, and each individual aircraft is issued an airworthiness certificate based on the type certificate. The FAA is even supplied office space on site by the manufacturer and many of its independent suppliers for ease of oversight and certainty of control.

ATR aircraft are inspected and approved by an agency of the French government, the Director Generale de l'Aviation Civile (DGAC). After being certificated in France, the airliner passes into the United States with a minimum of oversight and review under the Bilateral Agreement, a special trade treaty. The FAA process rubber-stamps the French certificate and then issues the U.S. type certificate. The FAA relies almost entirely on the data prepared by the manufacturer and reviewed by the DGAC and does little, if any, independent testing or monitoring. FAA test pilots spent about 10 hours in the ATR-42 to review cockpit layout and perform routine flight maneuvers.

The Bilateral Agreement was originally signed on August 9, 1973, and updated on September 26, 1973. Both the French and Italian governments are party to the Bilateral Agreement that, in effect, provides the law under which the United States and the cosigners of the treaty share responsibility for the certification of airplanes.

Several terms define the responsibilities of each party that make the treaty work. The official definitions under the agreement may vary slightly from those explained here; however, the following information clarifies a legally complex situation with a simple organizational overview and project flow.

The certificate holder is the organization that is granted the authority to manufacture the aircraft. These certificates have a value and are often later sold or licensed to other parties. In one of the last of many requests for special dispensation sought by ATR during the certification process, the consortium sought a dual-party certificate, listing both Aerospatiale and Alenia as certificate holders. The FAA had concerns regarding this request, because it could muddy legal issues pertaining to official notification and the identification of the controlling authority. After some internal discussions, the agency relented and issued the dual-party authorization with the understanding that only the French final manufactured version would be marketed in the United States under the certificate and the French DGAC would retain its role as the responsible oversight agency. It was originally stated that Alenia may seek authorization for an Italian manufactured version under the auspices of the Registro Aeronautico Italiano, the Italian counterpart to the FAA and DGAC; however, apparently this plan never reached fruition, and all ATR aircraft are of French origin.

For the ATRs, the DGAC has the responsibilities under the Bilateral Agreement as the certificating authority, which gives it oversight regarding the airworthiness of the aircraft. The ATR-42 was the first foreign aircraft certificated under the Bilateral Agreement and Joint Airworthiness Regulation (JAR) 25, which closely parallels but does not necessarily duplicate FAR Part 25 of the United States, under which Boeing and McDonnell Douglas jets receive type certificates.

With the DGAC acting as the certificating authority, the FAA acts as the validating authority. Under the Bilateral Agreement, the FAA is to be kept apprised of the progress of the venture through its office in Brussels, which is the contact point for the project. Specialists from the Aircraft Certification Office of the FAA in Seattle were involved in the work at various points and maintained overall control in the United States until the aircraft became certificated. After certification was completed and normal airline operations began, control passed to the Aircraft Evaluations Group (AEG) of the FAA, which monitored for continued airworthiness. This convoluted overlapping of sub-agencies reduced or eliminated individual responsibility on the ATR project and is endemic of the bloated bureaucracy at the FAA.

The entire certification process began with ATR making application to the FAA, advising the agency of its intent to design and manufacture a new aircraft and subsequently market the airplane in the United States. A meeting was initially held in Seattle to discuss with all parties the policies, practices, and standards of the FAA. The process was handled by the French DGAC with involvement by the FAA limited to the issuance, and hopeful resolution, of issue papers to DGAC and ATR to relay developmental and design concerns and information regarding the process. For the ATR-42, 98 individual issue papers were published during the project.

One of these issue papers, SE-9, written by FAA personnel, mentioned ongoing concerns regarding the ice-protection system on the ATR-42. With several of the concerns of the FAA still unresolved, especially those relating to the ice testing and the aircraft's ice-protection systems, ATR and the DGAC came forward with documentation about two weeks before the U.S. certification date of October 25, 1985. Unfortunately for ATR, the documents raised more questions than they answered in the mind of FAA aerospace engineer Frank vanLeynseele.

Piling a mountain of paperwork onto government officials shortly before a deadline is a well-known business tactic used by those who hope the overworked public employee will be overwhelmed by the review and simply rubber-stamp the data. Frank vanLeynseele does not use a rubber stamp.

In a memo sent to FAA officials on October 10, 1985, vanLeynseele stated, "A number of reports requested by Issue Paper SE-9 were received yesterday. In view of the short time remaining until TC (Type Certification), you should be aware of Systems and Operational concerns. The concerns expressed address airframe effects only."

A portion of the memo stated:

The Flying Qualities and Performance Evaluation of the ATR-42 subjected to the effects of natural icing and ice shapes suggest degradation which prompted flight manual restrictions and stall warning system modifications. . . . The pilot will be burdened with an additional workload. He must be alert at all times to detect the presence and severity of ice and changing conditions [that] require him to apply the proper corrections.

In layman's terms, with the degraded performance and handling characteristics, the ATR required special procedures be applied during operations in icing conditions and with ice on the airframe. This cure flew in the face of the long-standing FAA policy that maintains "long-term operational safety is better assured by design changes to remove the source of the potential problem, rather than by relying on long-term operational limitations."

Engineers and systems designers have long understood the fallibility of humans, especially when paired with machines. A respected aeronautical engineer and retired airline pilot recently told me, "If successful operation of an aircraft is hinged on the pilot completing a set of tasks, no matter how simple or limited, in a precise, required order every single time, it is just a matter of time before you experience a catastrophic failure of the system." That concern is quadrupled when the human is asked to subjectively determine the parameters of that operation, as in icing conditions.

Frank vanLeynseele closed his memo by suggesting an FAA meeting be held to "discuss the topics and decide on their acceptability." A disagreement over the icing performance on the ATR-42 had brewed for over a year. Icing certification was one of the issues for which ATR had consistently sought exemptions and offered numerous qualifications to the testing.

The FAA issued the U.S. type certificate to the ATR-42 on October 25, 1985. The journey that ended in a soggy bean field in northwest Indiana nine years and six days later had begun.

On June 6, 1986, an ATR-42 operated by the manufacturer suffered what the DGAC classified as a "temporary loss of control . . . resulting in a large negative attitude upset" during a crew training flight in instrument meteorological conditions. The aircraft was operating at an altitude of 14,000 feet, and the crew observed ice accretion on the wing. While slowed to 120 knots of indicated airspeed in a configuration using 45 degrees of flap and the landing gear in the down-and-locked position, the crew performed a "push over" maneuver. The result was severe pitch oscillations and altitude loss. This crew had the aircraft configured for a landing stall-and-recovery procedure practice, a procedure designed to train crews for an unexpected loss of lift while maneuvering slowly and close to the ground, i.e., while landing.

This loss of control was not unexpected. In the course of development and certification flight tests, the ATR-42 was reported by the French DGAC as having adequate handling qualities and normal stick (control column) forces in all conditions except with icing on the tailplane and flaps at 45 degrees. The manufacturer had prohibited the use of flaps at 45 with ice accretion for this reason by providing operators with a limitation in the aircraft flight manual. Documentation on this incident did not reveal whether the crew noticed the ice accretion before or after the upset and recovery. If they noticed the ice before the maneuver, why would the crew, especially factory personnel, attempt a flaps-45 stall when they had access to the test data and knew the potential problems? If the crew only noticed the ice after the maneuver, then the certification of the aircraft without an ice-detection system may have been in error and provided an early warning to the manufacturer.

According to an American Eagle document pertaining to this early loss-of-control incident, the "push over" maneuver was in conjunction with practicing of a stall with the flaps at 45 degrees, which was a potential exposure scenario in early ATR operations in "normal revenue service" because landings with the flaps at 45 degrees were routinely accomplished. How much altitude was lost in the French incident is not known, but 14,000 feet is much more height than a landing ATR would have when maneuvering with flaps at 45 degrees should the crew have been confronted with a tail stall problem.

After the June 6, 1986, incident, the FAA prohibited the use of flaps at 45 degrees in the ATR-42 for all operations except emergencies and, in an airworthiness directive, called for a physical blocking device to be placed on the flap selection handle. A July 24, 1986, FAA memorandum enunciated the positions of the agency. The memo reads, "The reason the FAA originally accepted the ATR-42 flaps 45 limitation [operation with flaps 45 only with no ice accretion] for U.S. type certification was because of the request made by DGAC and Aerospatiale, since this limitation covered substantial performance reduction and not handling qualities." At least this time, the FAA was true to its policy and even stated so in the memo: "FAA policy is to disallow airplane configuration limitations to cover handling quality deficiencies after an icing encounter."

ATR designed the ATR-42 with a position for flaps at 45 degrees to improve landing performance. It was properly assumed by the manufacturer that some need existed for aircraft operators to use short landing fields and steep approach angles to some airports. The flaps 45 position would help them with performance on short airfields and compliance with FAA-mandated operational criteria.

In the end, ATR, with DGAC concurrence, labeled the final investigative cause of the incident as "improper stick control." This conclusion placed the blame squarely on the shoulders of the flight crew and, along with the compliance to the airworthiness directive, almost closed the matter with the FAA.

When the prohibition was issued against the use of flaps at 45 degrees on the ATR, however, it hit a Hawaiian airline hard and would have precluded, or severely limited, the operation of the ATR-42 at some of its airports. The airline petitioned the FAA through ATR to lift the ban for operations in the Hawaiian islands, and after some discussion, the FAA relented and authorized the use of flaps-45 during limited operations. In their paperwork discussions occurring in mid-1986, the involved FAA personnel exhibited some understanding and foresight when they stated, "ATR-42 airplanes are known to be sensitive to icing conditions."

U.S. certification standards under Part 25, the transport aircraft regulations for aircraft with 20 or more seats, require that pilots be able to see ice on the airplane in their normal forward scan. The center window post, windshield wipers, or other protrusions that are in the forward vision of the crew have been allowed to meet this requirement on other aircraft. I have flown the ATR for more than 3700 hours and have rarely seen ice accumulation on the center post of the windows before a significant buildup. In addition, I am 6 feet 4 inches tall and usually have my seat in a full aft and just above the full low position, and I cannot see the windshield wipers from there without raising the seat. On the early models of the aircraft, no other protrusions were visible to the crew in the forward scan, but the FAA passed over this important operational requirement.

ATR's own test results document states under the procedures section, "The flight tests in natural icing conditions have shown that it is not necessary to fit the ATR-42 with a specific icing indicator. The observation from the flight compartment [cockpit] of the wing leading edge and spinners provides the crew with an unambiguous visual clue for the activation of the pneumatic de-icing system." The problem with this conclusion is that the pilots must look out the side windows and backward at the wing to view these items. Certification proceeded unhindered with this conflict to regulations.

The early detection of ice accretion is important; pilots must always be aware of the changing atmospheric conditions, particularly ice, to make proper decisions. Recognizing the potentially weakest link in the system, the regulations also refer to an "inattentive pilot" and "delayed pilot action" when developing the rules for ice certification. The deice equipment must function properly even if the pilot is inattentive enough to allow 1 inch of ice accretion on the aircraft before activation of the system.

On August 23, 1985, ATR reported in a document titled "ATR-42 Ice Certification—Summary of Flight Tests Results" that testing indicated their system performed with ¾ inch of ice accretion, "thus demonstrating the tolerance of the system with regard to a delayed pilot action." In the same document, the manufacturer added this concluding statement: "Would the pilot be inattentive, the hammering of the fuselage due to ice shedding from the propellers is considered as an unmistakable warning." The FAA allowed the ATR to be certificated without objecting to these deviations from the standards. In more than five years of flying the aircraft out of Chicago O'Hare, I have never heard ice shed from an ATR propeller.

The first U.S. operators of the ATR-42, including Simmons Airlines, sent some of their pilots to France for training in the operation of the aircraft, mainly because of the availability of aircraft and qualified instructors. This program should have allowed for the most in-depth and thorough training possible, because the designer and manufacturer of the aircraft were responsible for the course content and results. Many of those initially trained in France by ATR returned to the United States to train other pilots, initially with the assistance of ATR instructors, and later on through FAA-approved and FAA-monitored programs.

With Simmons slated to receive its first ATR in mid-1986, pilots were sent to Toulouse for ground and flight training, and by the fall of 1986, Simmons was operating four ATR-42s. On December 18, 1986, Simmons had the first of what would be numerous icing-related control problems. The aircraft involved were the third and fourth received, N422MQ and N423MQ, and the incidents occurred within 80 minutes of each other on an overcast morning at Detroit, Michigan.

At 8:55 A.M., Simmons Airlines Flight 2860 was approaching Detroit Metropolitan-Wayne County Airport. Captain Robert Fries was in command, Charles Long was first officer, and Lynn Venturio was the flight attendant on N423MQ. The flight, which had 35 passengers, was progressing normally from Traverse City, Michigan, and was in radio contact with approach controllers. The crew was told to expect an instrument landing system (ILS) approach to runway 3R and was receiving radar vectors to the approach course.

The weather and airport information at Detroit-Wayne was reported by the Automatic Terminal Information Service (ATIS) as information "echo." The crew recorded the data as sky overcast at 1500 feet above the ground with 10 miles visibility; temperature was 34 degrees Fahrenheit with a dewpoint of 29 degrees; the winds were out of the northwest from 300 degrees at 14 knots. The barometer was 29.83 inches of mercury, and parallel approaches were in use to both runways 3L and 3R. An advisory was provided for moderate mixed icing in the clouds, and the top of the overcast was 4900 feet above the ground. The clouds the flight needed to penetrate in the descent were about 3400 feet thick.

Flight 2860 entered the tops of the clouds as it passed to the east of the airport during its descent to 4000 feet and immediately began accreting ice. The crew had the required anti-ice systems operating and had discussed their plan of inflating the deice boots when they broke out of the overcast. The captain described the ice accretion as relatively rapid and the ice as rime. Approach control instructed the aircraft to slow to 170 knots and cleared the flight for the approach. The crew was to call the tower for landing clearance on passing the Huron outer marker, 5.5 miles from the runway.

The flight crew had finished the descent and approach checklists and was monitoring the autopilot. The aircraft had been in the clouds and picking up ice for about four or five minutes and was leaving 4000 feet to start its final descent. The first officer was the flying pilot with the autopilot on.

8:56:59: "I'll capture the glideslope and follow it down," the first officer informed the captain.

8:57:03: "Yeah. Lights are coming on. Yeah, we are icing up."

8:57:41: "Nothing like the Short; I'm surprised them guys can hold 150 in this." Captain Fries was talking about the rapid ice formation and the fact that the Shorts 360 aircraft he had previously flown would not be able to maintain 150 knots of airspeed with this type of accretion. "I doubt if you'll have to pull that power that hard to slow down."

8:58:05: "Uh, uh. This thing will come right back." The drag had built on the aircraft, forcing the crew to use extra power just to maintain its airspeed on the glideslope.

8:58:06: "Right out of the damn sky—so damn heavy."

8:58:15: "Do you want to pop it on short final?" First Officer Long queried, to verify the previous plan of inflating the deice boots after the aircraft broke out of the clouds.

8:58:17: "Yeah. The way it's building up, you know what I mean, it's not a slow buildup, it's fast. And we're going to pop out at 1500 . . . 300 to go to the marker." Captain Fries made reference to the reported height of the cloud deck where they anticipated breaking out into the clear and indicated that the aircraft was 300 feet above the altitude at which they would cross the Huron outer marker on the glideslope.

8:58:32: "That's right."

8:58:35: "And that checks. And there's Huron, that's showing 2700, so everything checks. There's Huron." The captain was verifying the navigation for his copilot and then made the call to the tower. The flight was 5.5 miles from the runway, still in the clouds, and about 2000 feet above the ground. "Simmons 2860's Huron inbound."

8:58:51: "Simmons 2860, Metro Tower, traffic short final. Cleared to land three right. Wind three zero zero at eight."

The first officer reduced engine power to slow the aircraft from the approach speed of 170 knots to lower the flaps and landing gear. Maximum speed for operations with 15 degrees of approach flaps was 160 knots (later increased to 170 knots) and the operating speed for the landing gear was 160 knots. To avoid any overspeed warnings, which even this fairly new crew knew would sound down to near 155 knots, the first officer waited until the aircraft slowed to 155 knots and then called for flaps.

8:58:57: "Flaps fifteen."

8:58:58: "Cleared to land, three right," the captain confirmed before complying with the flying pilot's request for flaps. He selected the deice boots while speaking on the radio.

At that second, the autopilot disengaged, and the new ATR-42 began to shudder and rolled sharply to the right, stopping the roll at about 40 to 45 degrees of bank and nose down to between 10 and 30 degrees below the horizon. First Officer Long grabbed for the control wheel and attempted a recovery. To his shock, the aircraft did not respond to the controls and oscillated in bank from 40 degrees right-wing-down to 40 degrees left-wing-down, descending rapidly.

8:58:58: "Oh!" an unidentified, shocked voice cried out as the autopilot disconnect warning sound was heard on the cockpit area microphone.

8:58:59: "Add power!" the captain commanded. "Add power!"

The aircraft continued for three alternate rolls to the left and right and the nose continued to fall as both pilots wrestled with the control column and Captain Fries advanced the throttles to the firewall. Flight 2860 began to accelerate, and the captain took control of the belligerent airliner from his first officer.

8:59:05: "Got it, wait . . . oh, oh, oh, oh, oh, oh."

8:59:08: "Okay!" Captain Fries exclaimed as the aircraft continued plummeting toward the earth.

8:59:09: "Pop the boots!" the first officer called, not realizing in his panic the captain had already set the airframe deicing system on.

8:59:12 to 8:59:16: The ground proximity warning system sounded in a mechanical voice that is purposely recorded without emotion so as not to add stress to the situation. "Terrain, Terrain. Whoop, Whoop. Sink Rate!"

8:59:18: "Oh, God!" The ground was rushing up toward the airliner, and in its severe nose-down attitude, the windshield was full of the Michigan landscape.

8:59:18: "Get the power, get the power!" Captain Fries commanded. The aircraft had accelerated to 190 knots and control authority was returning. Flight 2680 had fallen 500 feet in less than 20 seconds; recovery was accomplished just 1000 feet above the ground.

The airliner was less than 1000 feet above the ground and less than three miles from the airport. The runway was in sight, and the

ATR-42 seemed to be under control at 190 knots. Captain Fries elected to make a landing from this position rather than executing a missed approach and transiting the cloud deck another time. Because he could not lower flaps or gear at that speed without exceeding a limitation, he slowed the aircraft cautiously. At 160 knots, the captain called for flaps 15 degrees, and the familiar pitching of the airplane indicated the flaps moving into position and ensured that he still had adequate control. The gear locked down with a reassuring thud and the aircraft was landed at a speed higher than normal, near the 145-knot maximum for the 30 degrees of flaps used.

After the passengers deplaned and moved into the terminal, the pilots exited the cockpit and slowly walked around the aircraft to inspect the remaining ice formations. About one-half inch of rime ice remained on the unprotected surfaces of the ATR. The crew of Flight 2860 had dodged a bullet, and the crew lounge was buzzing with the story as it was passed from pilot to pilot in hushed but animated conversations.

The problem for the passengers and crew of Simmons Flight 2801 from Lansing, Michigan, to Detroit was that they were not privy to these discussions. At 10:15 A.M., Flight 2801 arrived in the Detroit area and prepared for an ILS approach to runway 3L. The weather had not changed from the time of arrival of Flight 2860 one hour and 20 minutes earlier. Unlike the earlier flight, 2801 flew only two to three minutes in the icing conditions and accumulated an estimated one-quarter to one-half inch of mixed icing by the time it exited the overcast.

The autopilot was flying the airplane with First Officer Steve Jobe acting as the flying pilot. Captain Joel Duran and flight attendant Michelle Erickson were the rest of the crew, and the flight carried 15 passengers. Michelle had noticed some ice on the airplane during her final walk-through but had not paid much attention to it.

On the ILS approach for most of the descent from 6000 feet, the aircraft had passed the Revup locator outer marker, and the crew had the flaps at 30 degrees and the landing gear down. The aircraft was ready for landing when Captain Duran selected the deice boots to remove the minor accumulation. Almost at the same time, the autopilot disconnected, and the aircraft rolled rapidly to the left, attaining a maximum bank angle of 30 degrees before rolling just as quickly to 20 degrees right-wing-down. First Officer Jobe fought the lightly loaded airliner through three oscillations before gaining control and making an uneventful landing.

At the gate, the first officer inspected the remaining ice on the aircraft, noting one-quarter to three-eighths of an inch on the unprotected surfaces. For the second time in one day, an ATR with only minor ice accretion had suffered departures from controlled flight.

None of the six crewmembers reported a buildup anywhere near the 1 inch anticipated to be collected by an aircraft with an inattentive pilot during certification work. Yet both aircraft experienced roll oscillations coincidental with the activation of the deicing system while in different configurations.

An immediate investigation was launched by the airline. Pilots filed reports with the company, which were reviewed and forwarded to the FAA. In a later safety report filed with NASA, Captain Bob Fries made an insightful comment regarding the control anomaly. Although he did not realize it at the time, Captain Fries touched on an issue that some experts later contended might have hindered or prevented the Roselawn crew from regaining aircraft control. He wrote, "My opinion is the aircraft was in a Dutch roll situation that increased to a stall, possibly aggravated by the spoilers." A Dutch roll is a phased, oscillating roll-and-yaw motion of the aircraft and is a negative quality related to the inherent stability of the craft.

The ATR aircraft use conventional ailerons for roll control but also use a hydraulically actuated spoiler on the upper surface of each wing to assist the ailerons at low speeds. The spoilers are operated by hydraulic system pressure; however, their deployment is controlled by the degree of displacement of the aileron. Although the ailerons affect the aircraft along the longitudinal axis (from nose to tail) by affecting the angle of attack of the wings and increasing or decreasing lift to their associated wing, spoilers actually "spoil" the lift of the wing by protruding into the airstream and disrupting the airflow. On the ATR, a slight time lag occurs between rapid aileron deflection and actuation of the spoiler, and in later investigation of an Italian accident in October 1987, some experts suggested the aircraft design might have exacerbated the roll excursions by placing the corrective inputs by the pilots slightly out of phase with the oscillations. If this is true, the attempts of the pilots to regain control of the aircraft might have actually delayed or prevented the resumption of controlled flight. In any discussion, however, experts seemed to agree this design decision of tying the spoiler actuation to aileron deflection accelerated the rapidity of the rolls.

Both captains of the Detroit aircraft made statements in their NASA safety reports with regard to the training on the deicing system performed by ATR personnel both in Michigan and France. The normal operation of inflatable boot deicing systems is to wait for the accretion of a moderate amount of ice before selecting an inflation cycle of the boots to crack the ice, which then is blown off the aircraft by the airstream. Simmons Airlines confirmed the pilots were never taught any different technique during their factory or company training sessions. After the incidents, ATR issued changes to its operating procedure manuals, and Simmons retrained its pilots to operate the deice boots constantly during ice accretion.

Possibly because only six ATRs were in operation in the United States at the time, the FAA erred on the side of safety and reacted quickly by issuing an emergency airworthiness directive on December 19, 1986, to prohibit the ATR-42 from operating in known or forecast icing conditions during airline operations until a cause could be determined for the December 18th events. The DGAC protested the action by sending a telex message on December 22. Their slight delay in responding to the grounding resulted from the lack of government staff over the weekend; the airworthiness directive was issued on Friday. When the telex did arrive, however, the nature of the message did not concern flight safety but rather what the DGAC stated as the erroneous reasoning for the airworthiness directive that in its view "gives an unbalanced view of known facts." The communication chastised the FAA for not complying with the "spirit of the Bilateral Agreement" by not consulting with the DGAC before taking urgent measures to protect the U.S. flying public. The French ambassador in Washington, D.C., also received a copy. Adding insult and a diplomatic threat to the message, Frantzen of the DGAC wrote, "Could you provide reference of any Telex or telephone for such consultation?"

In a campaign that was repeated often between late 1986 and 1995, ATR and DGAC mobilized an offensive that was aimed at quickly clearing the reputation of the airplane with data from the manufacturer and political clout. Before the U.S. investigation of the Detroit incidents was underway, ATR had sent a delegation to the Seattle FAA offices to discuss the matter and lobby for the aircraft. With amazing speed for such a complex issue, ATR proposed a fix to the problem on December 23, 1986, which the FAA officially accepted in its final rule adopted on June 26, 1987, effectively closing the issue.

The reclassification of the deice system as an anti-ice system and subsequent change to the use of the boots, which would now be operated continuously in conditions conducive to the formation of ice, caused new problems. Testing later revealed that this increased operational frequency of the devices reduced the boot life by a factor of six. This reduction in component life caused significant increases in maintenance costs to the airlines, and ATR again changed its position in mid-1987 by requesting still another means of compliance with the concerns of the FAA.

This alternate means of compliance, which the FAA deemed to meet an acceptable level of safety, removed the sensing of the stall threshold from the activation of the airframe deicing boots and placed it under the control of an anti-ice function. This anti-ice device is operated whenever the aircraft is operated in an icing atmosphere, which is defined as temperatures below 40 degrees Fahrenheit and the presence of visible moisture, such as clouds, fog, or precipitation. With this change, the boots returned to being a deicing device; however, they were to be operated "as soon as and as long as ice accretion develops on [the] airframe" from the previous procedure of "when ice accretion is observed." The operating procedure was still in conflict with normal pilot experience but followed closely the precertification testing.

ATR also changed the minimum climb speeds as a result of the Detroit incidents. Originally, the aircraft was allowed to climb at 130 knots with no ice accretion and 160 knots with observed icing. After the investigation of the Simmons events, ATR discontinued the use of the 130-knot speed, believing pilots could mistakenly use an improper, dangerously slow speed with undetected ice on the aircraft.

Another regional airline experienced similar problems weeks before the December 18, 1986, incidents, but that airline, the manufacturer, and the FAA did not share any details with other operators until after the Simmons icing events. In the end, ATR dismissed the Detroit incidents as pilot error for flying too slowly and improperly delaying the selection of the deicing boots. Although this conclusion contradicted facts, the manufacturer established a position that problems were never the result of the aircraft but were caused by human error. During later investigations of incidents, ATR created the presence of turbulence when none was reported to solve an incident without injuring the reputation of the aircraft. This tactic was used to "solve" the icing incident Dan Rodts and I experienced in Marquette in 1993, despite the fact that we encountered no turbulence.

Less than 10 months after the Detroit incidents, two ATR-42s operated by the Italian airline Aero Transporti Italiani, or ATI, the regional arm of Alitalia, had control difficulties in ice, oddly enough, also on the same day. ATI had seven ATR-42 aircraft, only one of which had been modified with the lower stall warning threshold tied to the aileron horns as required by the U.S. airworthiness directive. On this night, one of the crews was not as quick or as good or as lucky as the other. The ATR again demonstrated its weakness in ice and the political power wielded by the manufacturer.

On October 15, 1987, a weather system over northern Italy caused widespread rain on the ground, snow in the mountains, and ice problems for aircraft operations. Several aircraft reported moderate icing conditions in the area that day, and a few smaller aircraft diverted to another airport. Italian meteorology reports to the pilots of both ATI flights forecast and indicated the presence of moderate ice for their flight routes; however, nothing reported to the crews before departure indicated any weather phenomena that would preclude flying.

ATI Flight 460, an ATR-42, carrying 34 passengers and a crew of three, readied for departure from Milan to Cologne, Germany. The required route took the flight over the Alps and into the teeth of the storm system. Even though the sky had darkened several hours earlier, the aircraft departed Milan at 7:13 P.M.

The crew had been properly trained and operated the aircraft according to the ATI company operations manual. Updates of ATR operational policies issued during the spring of 1987 in the United States as a result of deficiencies uncovered after the Detroit incidents were not issued to the pilots of ATI Flight 460, and the captain opted for a climb speed of 133 knots, which had previously been an acceptable procedure. This night, it was a fatal mistake.

Unknown to the crew of Flight 460, four other aircraft were having some difficulties with structural icing over northern Italy. A King Air business aircraft experienced rapid ice accretion and some control problems that were quickly overcome. A DC-3 diverted its course after an encounter with severe ice. ATI Flight 1448, another ATR operated by ATI, experienced undetected ice accretion with the deice boots off, and the aircraft stalled. The crew recovered the airliner successfully and landed without further incident.

Flight 460 continued its climb at 133 to 135 knots for nearly 15 minutes without incident, and the autopilot was controlling the aircraft exceptionally well. The crew might have confused the climb speeds to be used for this particular flight condition, because the operational speeds of the ATR at that time were variable with conditions as observed by the pilots and the subject of a number of changes since the delivery of the initial aircraft. Note that the *observation* of ice was the controlling factor in the early years of ATR operation. Later, the speeds were predicated on operation in icing *conditions*, regardless of the status of accretion. Operating speeds also significantly increased after each control incident. Therefore, at this point in the evening of October 15, 1987, the minimum speed during the climb should have been 135 knots if the crew did not observe icing and 145 knots if they did. According to subsequent reports, the crew was operating the deice boots at some point in the flight; however, it is not known whether ice accretion was observed or whether the crew had operated the boots to determine whether any ice was masking itself as a wet-appearing glaze.

At 7:26 P.M., the autopilot began sensing a loss of performance and adequate lift. Small perturbations of the ailerons were detected for several seconds on the digital flight data recorder output as attempted compensation for a developing instability was sought by the computer. To the pilots and passengers of Flight 460, this would have been detected as a barely perceptible rolling motion. The flight crew also would have noticed a minor rocking of the control wheels and a possible oscillating roll on the electronic attitude display indicator. The minor movements were probably discounted by the crew as the result of a number of possibilities.

At 7:26:42 P.M., as the flight was climbing through 16,000 feet, the autopilot disengaged, and the aircraft rolled rapidly to 45 degrees and then the roll immediately reversed. The nose of the aircraft fell as the roll oscillations quickly diverged, reaching maximum amplitude of 120 degrees. Rarely during these rolls was the aircraft ever in a steady state. As soon as the roll reached a maximum bank in one direction, it began in the opposite direction. The crew suddenly was forced to deal with a wildly rolling aircraft that was achieving greater and greater divergence from level flight without any visual cues from outside the aircraft.

The pilots fought with the craft, at times with both crewmembers attempting to operate the controls. They had no time to worry about the passengers and flight attendant who, at that time, were being thrown from side to side of the aircraft as it began to fall from the murky sky. Screams erupted from the cabin, but these screams went unheard by the captain and first officer. The control column was pulsating and vibrating as the shocked pilots attempted the impossible.

Aeronautical engineers can predict the flight characteristics of a clean wing with a fair degree of reliability; however, when ice is added to the equation, airfoil behavior is impossible to guess. This is especially true when the airfoil is at, or attempting to recover from, an excursion near or beyond the stall threshold, as was the case of Flight 460. With nothing but flight instruments and experience to save them from a disorienting and violent situation, the pilots had little time to regain their composure, sort out the information, and correlate it to an appropriate recovery technique.

Evidence exists of significant amounts of ice accumulation on the tail surfaces and wings. Elevator movements also puzzled investigators as the initial fluctuations changed to a nearly constant full nose-up deflection until impact. In addition, the pitch disconnect mechanism, which was placed into the control circuitry for protection of a blocked elevator problem, had activated, probably the result of the action of the pilots in fighting the three firings of the stick pusher. The captain ended up flying the left side of the horizontal tail and the first officer had control of only the right side.

Cockpit voice recorder transcripts revealed the pilots complaining of the controls being "blocked." The rolling, pitch disconnect, and reversals of control forces created confusion. The aircraft did not respond to pilot inputs. Once the speed passed 350 knots, the controls might have been aerodynamically locked, precluding pilot movement, because of the high-speed airflow. Regardless, it was too late.

The fate of Flight 460 was sealed eight seconds before the autopilot disconnected, when the pilots selected a slight decrease in the climb speed. The fall to earth took 45 seconds. Photos of the crash site were similar to those taken of the remains of Flight 4184 seven years later. Little of the aircraft or its occupants remained intact when the ATR shredded its way through tall trees and struck the rocky ground at high speed at the 2300-foot level of Mount Crezzo near Lake Como in Northern Italy.

A short distance away, at the Modonnina Refuge, an inn in the vicinity of Mount Crezzo, the proprietor, Franco Villa, was eating dinner when he heard a noise and ran to the window. He reported seeing a ball of fire on the side of the mountain, followed by the sound of an explosion. Police and rescue workers searched the area throughout the night, but the wreckage was only spotted at first light the next day. Rescue workers arrived by helicopter and immediately returned to their base, reporting that there was nothing to be done. Recovery workers returned and began the operation of removing the bodies and wreckage.

Surprisingly, the story of the "Italian accident," as it became known in United States investigative and regulatory circles, was not the crash itself but the after-effects of the tragedy. Within a few hours of the accident, Mario Del Franco, a local prosecutor from Como, arrived on the scene and began an investigation under Italian law. Del Franco assembled experts from around the world to work on the team and, especially, to examine the ice-protection systems of the ATR.

This battle was fought for years in conference rooms, board rooms, hearing rooms, back rooms, and finally court rooms across the globe. Within one week of the accident, ATR was rejecting criticism of the ATR's deicing system, saying it worked normally and adding that the company intended to defend the quality of its products and its image, by legal means if necessary, against "grave and gratuitous judgments in the press on the reliability of the ATR-42."

In the United States, the FAA apparently did not monitor or follow up on the Italian accident investigation. David Hinson, administrator of the FAA, gave the impression and even made statements in 1994 and 1995 after the Roselawn accident, that the FAA "did not request a copy of the Italian report."

U.S. operators of the ATR were notified of the crash at Como by telex the next morning. Almost immediately, a rift developed between the French and Italian authorities, and the Italian Air Minister moved to ban all flights of ATR aircraft in the country. The remaining six ATI ATR-42 airliners sat on the ground while the investigation progressed.

A day after the accident, both data recorders were found and sent to the United Kingdom for transcription while recovery of the human remains and aircraft parts continued. The Italian government asked for American assistance, realizing that the vast amount of technical expertise available in the United States could augment its own resources. Working through the U.S. embassy staff in Rome, which included James Murphy, an FAA representative, Mario Del Franco began discussions for the desired assistance. Months passed before the response came.

The DGAC and ATR corporation responded rapidly to throw suspicion from the aircraft. With another unprecedented effort, the DGAC informed John Varoli of the FAA Aircraft Certification Office stationed at the U.S. embassy in Brussels on November 6, 1987, of its desire to discuss "various solutions" at an upcoming meeting in Paris. The Brussels office of the FAA had been the main contact point in the initial certification of the ATR-42 and would soon be preparing for the work on the new ATR-72.

FAA officials in the Seattle office of the Aircraft Certification Office watched the early activities with growing concern. Less than one year earlier, these engineers had a warning of the susceptibility of the ATR to ice-related problems. Now some in the FAA found the Italian crash disturbing. The Seattle office of the FAA made plans to send Gary Lium, an engineer and private pilot, and Howard Greene, an experienced senior FAA test pilot, to the ATR factory in Toulouse to reassess the performance of the ATR in ice. Their report on their trip in December 1987 yielded interesting findings and opinions.

Fifteen days after the accident, Leroy Keith, the manager of the FAA Aircraft Certification Division responsible for the oversight of aircraft manufactured outside the United States, prepared a briefing document on the ATR problems and FAA actions to date. By this date, four U.S. carriers had a combined 18 ATR-42 aircraft in service. Five operators were in the process of approval. A wave of ATRs was coming, and now seemed a good time to stop it. Ironically, the first icing-related airworthiness directive issued on the ATR and prompted by the French training incident was effective on October 15, 1986, exactly one year before the Italian accident.

On November 18, 1987, Leroy Keith sent a telex to Pat Lapasset, his DGAC counterpart. The message read, "We appreciate your keeping us informed regarding the ATR-42 accident in Italy. In light of our

Airworthiness Directive actions on this airplane and the RAI's [Registro Aeronautico Italiano] actions, we believe you will agree that a review of the ATR-42 icing certification would be in order." Leroy Keith's superiors later overruled his suggestion of a certification review by issuing a strongly worded memo that clearly supported the ATR.

When Howard Greene and Gary Lium traveled to Toulouse to review and test the ATR in icing conditions during December 1987, the Simmons incidents weighed heavily on their minds. Greene noted Simmons Flight 2860 "exhibited very similar characteristics to the initial phase of the [Italian] accident." According to recently released documents, his report on March 27, 1988, was "coordinated" with the French authorities, yet the coordination did not preclude Howard Greene from including some damning data.

Regarding the wing design, the report was cautious but complimentary of ATR's efforts. "The wing section is a modified NASA 65xxx series, modified to spread leading edge negative pressure peak out chordwise [from leading edge to trailing edge], i.e., laminar flow. Any discontinuity will trip it back to ordinary 65xxx characteristics, with consequent large drop in C_{LMAX}. It is surprising that ATR could make this section so effective with a boot installation." In other words, the wing was susceptible to contamination, that is, "discontinuity," with the result a large drop in performance. C_{LMAX} means the maximum coefficient of lift. The ATR wing relies on a smooth airflow, and any contamination, such as bugs or ice, causes large disruptions of the flow and significant performance degradation.

ATR accidentally opened the door to an interesting conflict when it showed the FAA team a group of photographs. In all previous conversations, and even in its own document "ATR-42 Ice Certification— Summary of Flight Tests Results" dated August 23, 1985, ATR stated, "In all encountered icing conditions the wing surface behind the boots was clear of ice. The chordwise extent of the boots resulting from the ice buildup analysis proved so to be correct." Making a tactical error that they obviously never intended, ATR showed Howard Green its own photographs of ice buildups in ridges *behind* the deicing boots.

Minimal flight testing with small ridges of ice shapes behind the boots was conducted during this trip but failed to reveal anything wrong. Greene was not entirely convinced, however, and his report conclusion section contained the statement, "Wing sensitivity to small amounts of contaminant has not been a factor in any of the incidents, yet."

On January 25, 1988, one month after Howard Green and Gary Lium had done their flight testing, ATR, the DGAC, and the Italian airline did some additional investigation. During one flight in ice, the plane rolled with uncommanded control movements that appeared to precisely duplicate what later occurred to Flight 4184. The information was withheld from the FAA by the French but was included in the Italian accident report, which nobody from the United States bothered to obtain.

The official Italian investigation into the accident near Como progressed slowly. Urgent requests came from the president of the Technical Investigation Commission for data in FAA files pertaining to icing problems with, and airworthiness directives issued for, the ATR-42 aircraft. Mario Del Franco commissioned a British government laboratory located at Bascombe-Downs, England, to perform ice testing on a full-size ATR, the results of which were furiously contested by the French. The French and Italian authorities battled over adjectives to be used in reports, and the U.S. government attempted diplomatic tightrope walking, not wanting to offend any of the three friendly nations. Original transcriptions of the cockpit voice recorder and digital flight data recorder were judged to contain "erroneous parameter readouts" and deemed unusable by the Italian magistrate, who then ordered them returned to the United Kingdom for a second transcription. By the spring of 1988, some conclusions were reached.

The testing at Bascombe-Downs involved subjecting an ATR-42 to blower tunnel icing cloud generation projected against the wing and horizontal tail and elevator of the aircraft. The engineers selected two airspeeds for the static testing, 133 knots and 145 knots; the first speed was that of ATI Flight 460 at the time of the upset, and the second was the manufacturer's recommended climb speed with ice accretion in 1987. A wide range of droplet diameter and liquid water content conditions were tested, with some tests involving a dual-rig to ensure the tail was adequately exposed to the generated ice cloud. The British understood the tests must include the tailplane airfoil because of its increased susceptibility to structural icing over that of the larger main wing.

During the late 1940s, the U.S. government conducted a study of the atmosphere to define an "envelope" of anticipated aircraft icing data to be used for aircraft certification for flight into known or forecast icing conditions. This envelope, known as "Appendix C," used the concepts of water droplet diameter as measured in microns and the liquid water content, or density of the cloud, as measured in grams of water per cubic meter. The certification standards use this definition of the atmosphere in determining the flight times in continuous exposure for stratus or layered clouds and intermittent exposure for cumulus or vertically developed clouds.

The chord of an airfoil is the distance through the cross section of the wing between the leading edge and the trailing edge of the wing. The impingement of water droplets onto the airfoil depends on several factors, one of which is droplet size, which relates to droplet mass. The greater the mass, the greater the inertia, which does not allow the droplet to be moved out of the path of the airfoil by the airstream. The impingement limit is the largest percentage of chord on which droplets impact and form ice.

The results of the Bascombe-Downs tests, although hotly contested, revealed some surprises that might have changed the course of events if investigated thoroughly. Of all the tests, the French came closest to accepting the results of Test 2, which used a droplet size of 40 microns with a liquid water content of 0.15 grams per cubic meter. These conditions are close to the extreme of Appendix C certification standards, and when applied to the ATR wing, the ice developed on 23 percent of the chord. A similar test criteria, when applied to the horizontal tail, covered 30 percent of the chord with ice, a percentage that was well beyond the limits of the deicing boots.

With the ATR-42, the limit of the effective wing boot—that portion of the black rubber bladder that inflates—is 5 percent of chord, and the ATR-72 limit of the effective wing boots is 7 percent of chord. After the Roselawn accident, ATR increased the effective coverage of the boot to 12.5 percent of chord on both aircraft, although this coverage was still smaller than the 23 percent impingement limit in the British tests and also less than the original commitment of 16 percent of chord made by ATR after the accident involving Flight 4184. The tail boot was not addressed in the retrofit and remains a concern to many experts.

The tests conducted in England were not the only issue over which the French and Italians battled. ATR officials came with the complete support of both the DGAC and the BEA, which is the French equivalent to the U.S. NTSB, and at every turn, the Italian investigation met with resistance when looking to the aircraft itself for fault. The Italians needed support, so they again asked the Americans for help. Internal FAA documents indicate that the agency was reluctant to become embroiled in an international incident, and the decision was made by senior officials of the Reagan administration to discreetly ask the Italians to make a formal, written request for assistance through the NTSB.

On May 3, 1988, the request came from the Italian Minister of Transport directly to James Burnett, then chairman of the NTSB. As directed by the FAA in Washington, the invitation was not addressed to the FAA. Indicative of the conflicts previously reported by monitors of heads of the FAA and NTSB, the letter stated, "During the investigations . . . some doubts and different interpretations have been raised about some icing tests made on wing and tail plane of ATR-42." The minister went on to detail his request: "In order to get further evaluations at an international level of absolute neutrality and professionality, I did decide to set an international commission of three members to be questioned about one or more problems. I would appreciate if you could designate, in the spirit of usual cooperation between our two countries, a member of the aforesaid commission having a sound experience as aeronautical engineer and test pilot preferably specialized in dealing with aerodynamic problems."

The Italians had made the required diplomatic move. It was up to the U.S. administration to respond. The NTSB responded by sending an electronic facsimile to James Murphy, the FAA representative in Rome, who then forwarded it to L. Alexander in the Economic Section of the State Department, also stationed in Rome. The response was sent on June 9, 1988, more than one month after the request from the minister. The NTSB International Liaison, Steve Corrie, in response to the request for an expert, stated, "Since that time we have been attempting to find an appropriate expert and have been trying to decide how to best respond to this request."

Corrie discussed efforts to seek out an expert and how they had even requested help of NASA. A later portion of the communique suggested some motivation for delay: "There have been enough conflicts already between the Italian and French authorities involved in the investigation, and we would not want to be in the middle of, or cause any further conflicts." The memo closes with a half-hearted attempt at an "interim measure" which will "hopefully give us more time to obtain the expertise needed."

If the original request had actually come on May 3, Corrie's response about needing more time to find a suitable expert might have been understandable; however, a priority message sent by the FAA in Washington to the Director General of Civil Aviation-Italy in early November 1987 stated, "It has come to our attention that Italian civil aviation authorities have invited FAA reps from the U.S. to participate in the investigation. We would like confirmation of this invitation."

An FAA memorandum from Albert Blackburn, the associate administrator for policy and international aviation, to the Rome FAA representative, dated July 29, 1988, shed more light on the FAA's stance:

> *Thanks for your comments concerning potential FAA involvement in the ATR-42 issue. You are right on the money about FAA not becoming embroiled in internal political and legal*

conflicts. I have discussed FAA's role in the ATR-42 icing issue with Craig Beard [the director of airworthiness at the FAA]. Craig and I agree that FAA should decline to provide comment on the proprietary Bascombe/Downs icing test. While we (and obviously, NTSB) will continue to avoid being thrust into any investigative or judicial review, we do need to maintain an appropriate technical dialogue with RAI. FAA's participation should be correctly focused on providing RAI with information and interpretations concerning certification standards. We do recognize that situations exist where FAA may be asked to walk a rather fine line between interpretation of certification standards in general and offering opinions on raw data. Per your observations, Craig and I agree that FAA should be extra cautious in our dealings with the Italians and French in this issue. We do not want to be caught in the middle, no matter how innocent our motives are. To this end, Craig and I will convey the concerns you raise, and which we agree fully, with our people.

This memo also overrode Leroy Keith's suggestion of a certification review of the ATR-42, which could have intercepted problems passed on to the ATR-72. Blackburn wrote, "Moreover, FAA has no plans to reevaluate its ATR-42 certification." Six years after this memo was sent, the passengers and crew of Flight 4184 were victims of the failure of their government to help fix the ATR control problems when it had the chance because it was afraid of irritating an ally with a diplomatic faux pas.

In August 1988, Italian officials met to discuss and review findings, admit new documentation into the docket, and prepare a final report, which was due in October 1988. According to the FAA, the October date was in doubt because numerous conflicts were still unresolved. The FAA representative in Rome attended the meeting as an observer. In his memorandum about the event, the representative noted that the French BEA attended the meeting as a member of the commission but held no voting rights. He also noted that the manufacturers were observers and were represented by the BEA. "It is interesting that the French BEA brought experts from the factory and were anything but unbiased in their opinions," he wrote. "It was very clear that their purposes [were] to [en]sure that the report says an absolute minimum pertaining to any suspected deficiencies in the aircraft."

Detailed discussions were held during the course of this meeting regarding a number of issues, mainly surrounding the subtle choices of language. The French did not want any references to factory operating manuals in the report and wanted only the consideration of data found in the airline's manuals. ATI pilots in attendance complained they had attended original ATR training at the factory and used factory-generated manuals that apparently indicated lower stall speeds than those currently in effect. After the Como accident, ATRs were grounded in Italy for three or four months, and on return to flight operations, the minimum operational speeds were significantly increased from preaccident values.

With regard to the Bascombe-Downs icing tests, all parties to the investigation agreed that the testing went beyond certification standards except for test 2, which showed 23 percent of chord accretion of ice on the wing and 30 percent on the horizontal tail. In the British study, the team concluded the deicing boots "coverage is insufficient to support aerodynamic flow over large areas of the wing and tailplane." ATR argued the ice encountered by ATI Flight 460 contained significantly larger droplet sizes than those envisioned by the certification standards, and even though Italian weather authorities classified the icing as moderate, ATR and DGAC had, almost since the date of the accident, classified the icing as severe or extremely severe.

Substantial industry concern had been developing since the late 1970s over the adequacy of the Appendix C icing envelope when compared with what really exists. Aviation experts and the NTSB had pleaded with the FAA for years before the Italian accident to review the icing certification standards so that they more closely resembled the exposures of actual operations. The FAA and NASA had a study of the atmosphere in place in 1988 with ongoing data acquisition and the promise of decisions to be made "next year."

After the retirement of Dick Adams, the National Resource Specialist for Icing, around 1987, according to the U.S. Government Accounting Office, the program wandered aimlessly. Even though research was completed, papers were written, and data distributed widely, FAA management did not take action or even take much notice. One of the reports commissioned by the agency, "A New Characterization of Supercooled Clouds Below 10,000 Feet AGL [above ground level]" was published in final form by the FAA in June 1983. The report stated that "this new characterization" should be used to establish design criteria and rules and regulations for ice-protection systems and equipment for aircraft that typically operate below 10,000 feet. The existing criteria applied to all aircraft seeking U.S. certification was promulgated in Federal Aviation Regulation (FAR) Part 25, Appendix C, and was deemed excessively conservative by experts. The research neglected to include flight in what was eventually called supercooled drizzle drops, although these same experts would later classify SCDD as the most likely exposure encountered by Eagle Flight 4184.

The Italian commission wanted to endorse the expansion of the Appendix C envelope and asked the FAA observer about the view of the agency on such a recommendation. In his report to Washington, James Murphy stated his response to the question was "They should make whatever recommendations they considered appropriate. I also gave them a copy of Craig Beard's letter dated May 25, 1988, which indicates we [FAA] are working on future revisions of the [Appendix C] curves." By late 1995, no action had been taken by the FAA on the modification of the icing certification envelope, other than more studies.

One of the final items of discussion was the use of proper adjectives. Because the Italian officials had classified conditions in the vicinity of the crash as moderate, and the ATR is designed to operate in moderate icing conditions, the French were demanding the use of the term severe to contend the aircraft was operating above design conditions, thereby eliminating any possible blame placed on the ATR design. Disagreement was lively, according to some in attendance. Although the commission relented and agreed to note in the report that seven aircraft diverted from the area that night because of icing, they would not commit on their final choice of descriptive language. The prospect of having the final report completed by October 1988 had an improbable prognosis on this day in August.

The DGAC responded in the summer of 1988 to NTSB inquiries of steps taken by ATR and DGAC to avoid further icing-related incidents: "Attached to this document you will find also various reports resulting from technical investigations carried on by the constructors but supervised by French authorities, which have led to demonstrate that ATR-42 certification is out of doubt, so confirming the preliminary conclusions established in November 1987."

The NTSB did not know at the time they received the DGAC response that ATR was struggling to explain another icing-related event. Just three months after the loss of ATI Flight 460, an ATR-42 training flight took off as ATI Flight 8860 on January 25, 1988. The aircraft carried three pilots, two qualified captains in the left and right seats and a first officer in the cockpit jumpseat. After three trips around the pattern, the crew performed touch-and-go landings and takeoffs. The crew then climbed to 5000 feet and performed a simulated engine failure. With all anti-ice and deice systems on, Flight 8860 then climbed to 10,000 feet to exit the tops of the overcast and began a series of practice stall exercises. The aircraft was about 1400 feet above the cloud deck in clear air. Even though they had operated the ice-protection system during the climb, both pilots made a visual inspection of the airframe, concurring in their observations of no ice accretion.

The first stall maneuver was performed with the ice-protection systems still operating, and the crew received the stall warning at appropriate speeds and completed a normal recovery of the aircraft. One minute later, the airplane was configured with 0 degrees of flaps, landing gear up. It is not clear if all anti-ice and deice systems were off; however, at least some were, which was appropriate with the lack of visible contamination.

As Captain Noseda flew the maneuvers by hand, he slowed the aircraft by removing power and holding altitude, with the result that the nose rose and the angle of attack of the wing increased. At 111 knots, the ailerons deflected to a right-wing-down position for a quarter of a second before the pilot was able to apply an opposing control force. The rapid deflection caused the aircraft to roll abruptly and unexpectedly to the right, reaching a maximum 26 degrees of bank in two seconds. Because he had been flying the aircraft by hand as required during practice stall maneuvers, Captain Noseda was able to begin a recovery immediately. As soon as the rate of roll appeared, the pilot started a recovery by pushing the control column forward, which lowered the nose of the aircraft and reduced the angle of attack, increased the power, and applied the counter force to the ailerons. In the appropriate response, Captain Plati, the nonflying pilot, placed the flaps to the 15-degree position. The short-lived event ended with the pilots regaining control, but not before the aircraft experienced rolling and control deflections.

With the crew still surprised from the roll, Captain Plati noticed ice detach from the aircraft. He estimated the thickness to be 2 millimeters (between one-thirteenth and one-twelfth of an inch). If Captain Plati's assessment was accurate, the statement in Howard Greene's report of "Wing sensitivity to small amounts of contaminant has not been a factor in any of the incidents, yet," may have been eerily prophetic. This control incident took place between the time Greene and Gary Lium made their investigations and observations in Toulouse and when Green wrote the final draft of his report containing that statement. Nobody, however, informed the FAA, NTSB, or other operators of the ATR. In fact, even after the Roselawn accident, the incident of ATI Flight 8860 was not included in the NTSB's chronology of control events. Considering the shocking similarity to other events and the occurrence during an ongoing investigation of a major accident of similar nature at the same airline, wider dissemination of the information should have occurred, unless, of course, the manufacturer was attempting to avoid further scrutiny of its product.

The mystery of ATI Flight 8860 was quickly put to rest when ATR completed insightful laboratory work on February 19, 1988. This work could be considered monumental as explanations offered at an earlier investigatory meeting held on January 27, 1988, were unverified, with ATR commenting as follows.

1. *After having analyzed the data of the Flight Data Recorder (FDR) the only explanation consistent with experience in natural or simulated icing conditions is that the roll departure could only be associated with a stall of the right wing. The angle of attack and Indicated Air Speed (IAS) at which it appeared are practically the same as demonstrated in test flight with 4 mm of ice accretion.*

2. *From the same experience mentioned above we have no explanation for the increment in aileron deflection from 2.3 to 4.6 degrees observed just before the roll departure. Our feeling is that even without this aileron deflection the stall would have occurred at same angle of attack or slightly higher.*

The ATR laboratory report seems to say that, although Flight 8860 had ice, no explanation exists for the uncommanded control movements and the aircraft would have performed normally had it not been for the ice. The report concluded that the crew was faced with erroneous temperature data. Apparently, according to the laboratory report, a large insect got into an internal channel of the external temperature probe, causing a reduction in the airflow around the sensing elements and subsequent erroneous data. Moving quickly to conclude the event, ATR drafted a memorandum that stated the probe was blocked, and they mandated a daily check of the probe because the probe is inaccurate if blocked. The matter of ATI Flight 8860 was closed. For the first time, the event was not blamed on the crew.

The final commission report on ATI Flight 460 was issued in December 1988, 14 months after the accident and subsequent to numerous confrontational discussions. However, Mario Del Franco, the local prosecutor, filed criminal charges against seven people, including Jean Rech, the ATR's French designer, on October 15, 1988, the one-year anniversary of the crash. The indictments charged the seven with multiple manslaughter and negligence and cited inadequate aircraft design and improper training and oversight by the airline. The legal wranglings eventually led to dismissal of the charges, refiling of new charges against an increased list of defendants, and three convictions and jail sentences, which would be overturned on appeal. Jean Rech was cleared of charges.

The final report on the accident cause cited improper flight procedures and training of the crew. Against the strong objections of the

Italian pilots' union, the flight crew was blamed for improper procedures, including inappropriate airspeed in the climb and failure to
use proper recovery techniques. In the end, the Italian commission
disregarded much of the Bascombe-Downs testing except for concurring on the limited efficacy of the boot's existing design and the
adoption of the British recommendation for the expansion of the area
covered by the deicing boots. The airline was found negligent in its
training and manuals. The wording of the final report classified the
cause as pilot error.

In my review of the available documents, including those translated from Italian reports, I found the conclusions were not entirely
supported by the facts, so I contacted several parties involved in the
after-accident issues to discuss matters. One possibility seems to have
been overlooked as causal or contributory to Eagle Flight 4184. The
report neglected to note the effects of the tail-ice contamination and
the role it played in the control movements shown on the data
recorder and the apparent confusion of the pilots when confronted
with improper and conflicting control forces and responses. Another
issue was the failure of the manufacturer to comply with the larger
boot recommended by many authorities. The recorded response of
ATR to those recommendations was "no." One other concern raised
but not satisfactorily addressed was the experts' opinion that the data
assembled did not support ATR's claim that the aircraft in the accident
had accumulated an abnormal amount of ice, "in excess of 3 inches."
This hypothesis of ATR was not supported in testing, and that conflict
was never explained. Finally, the ice ridge, later cited as "previously
unknown" during the Roselawn investigation, was identified and
stated as having the potential to "cause a severe degradation of the
aerodynamic characteristics." Although noted in some reports but not
fully addressed, the French airlines Air Littoral and Avianova reported
ice-related incidents; Air Littoral's incident occurred June 5, 1986,
Avianova's on February 24, 1988. The details of those incidents are
not well known, but they add two to the list of the ATR's ice-related
control incidents.

In discussions with C.O. Miller, a 40-year veteran of accident investigations and an independent consultant hired in 1987 to study the
Como accident, I commented on the political pressures evident in the
tone and origination of some documents. "The whole damn thing
was political and still is," Miller responded. His philosophy remains,
"every accident and incident is trying to tell us something; we just
have to be listening."

Even though the final report was toned down under pressure
from French officials, the document could have helped the FAA and

NTSB in their decision-making processes regarding the continuing airworthiness of the ATR. ATR could not completely explain the loss of control mechanism of the accident in Italy, and experts pointed to the ice accretion on the tail as a contributory factor in the inability of the crew to regain control. The reputation of the ATR was damaged with European airlines. The only change was the increase in airspeeds, forcing ATR pilots to fly faster with ice on the aircraft, which problematically increases the likelihood of tail-ice accumulation and possible pitch-control anomalies. Unfortunately, and for unexplained reasons despite the intense monitoring of the investigation by FAA and NTSB, U.S. authorities did not request a copy of the reports.

Just after the final report of the accident near Como, Captain Mike Bodak and First Officer Mike Monreal found themselves fighting an out-of-control ATR-42. On December 22, 1988, Simmons Airlines Flight 4295 was approaching the Central Wisconsin Airport near Mosinee, Wisconsin. The flight had originated in Chicago and had flown in cloud and moderate turbulence while en route at 16,000 feet, arriving in the terminal area shortly after 4:00 P.M. The aircraft, N427MQ, carried 34 passengers and a crew of three. N427MQ was the eighth ATR-42 delivered to Simmons.

Captain Bodak was flying the aircraft on autopilot with First Officer Monreal performing the duties of the nonflying pilot. The airliner was under the control of Minneapolis Center. Because of the distance from radar sites and the terrain elevations in the area, Minneapolis Center lost radar contact with the flight during its descent. Although the controller attempted to provide radar vectors to the approach course, the crew noted they would not intercept in an acceptable manner and subsequently requested a full approach flown on their own navigation. This approach would keep the aircraft in the rain and clouds for a longer time; however, the temperature at 6000 feet was well above freezing, and the crew was monitoring for icing. A temperature inversion was observed with the air at 6000 feet warmer than the air below, and rain was falling, which meant the possibility of encountering freezing rain as the aircraft descended on the approach. Weather at the airport just before the approach was reported as low overcast at 400 feet and visibility of 1 mile in fog with light rain falling. According to crew statements, all ice-protection systems were operational, and the pilots had discussed and raised their minimum speeds to match the flight conditions and the latest data from ATR.

The flight was west of the airport and maneuvering on autopilot to return to the approach course to use the ILS to runway 8. Captain Bodak reduced power in anticipation of lowering the first setting of flap. At 3000 feet, the flight was clearly in below-freezing temperatures, having crossed into freezing temperatures at about 4500 feet on the descent. The captain turned his autopilot heading selector for a right turn back toward the localizer, and the aircraft responded by entering a right banked turn of roughly 27 degrees.

"Hey, how are we doing on the ice?" Captain Bodak queried.

The first officer looked back over his shoulder at the wing and propeller spinner visible out of his rear side window and reported, "Negative ice, just wet. I'll make the call to Center to tell them we're turning inbound."

The aircraft decelerated to 155 knots and settled in at 153 knots with the reapplication of power. Unseen to the crew, the ailerons stopped responding in the turn and moved on their own to a nearly full left-wing-down deflection of minus 12.3 degrees. A minor vibration, which rapidly progressed for three seconds to become more violent, was felt by the crew. The autopilot disengaged, and the aircraft rolled rapidly out of the right bank and into an 80-degree left bank. Captain Bodak frantically grabbed for the control column and found he had absolutely no positive control of the roll axis. He advanced the power levers, and both engines surged to provide in excess of 120 percent of rated torque. Pitch oscillations joined the alternating high degree of roll as the aircraft began descending in a wing stall.

Fighting with the controls to no apparent avail and tossed by alternating G-forces peaking at plus 1.7 and minus 3.0, Mike Bodak began visualizing his little girl in her holiday outfit. "I could see her and that is all that my mind could focus on. I thought we were gonners because the airplane was not responding to my control inputs, we were going down, the ground proximity warning was going off 'terrain, terrain,' and then Mike Monreal hollered 'Props!' and I yelled back 'Max!'"

Captain Bodak relayed this to me when we flew together almost a year after the event. "The aircraft accelerated to about 190 knots and became manageable again. My heart felt like it was going to beat a hole out of my chest, and I just wanted to hold my daughter."

The crew regained control of the airliner 1200 feet above the ground on a heading that was taking them away from the airport. First Officer Monreal contacted Minneapolis Center to inform them of what had occurred and to ask for a vector back to the final fix. The flight was instructed to climb back to 6000 feet so Minneapolis Center could get adequate radar data. Center then aligned Flight 4295 for another approach. The crew added speed for the third approach attempt and then got the word from the tower of freezing rain at the airport a little earlier. Captain Bodak decided a continued approach to be the safest course and landed the aircraft on an ice-covered runway. After they parked the airplane, the pilots discussed their experience before leaving the plane. The following dialogue contains portions of their conversation, recorded on the cockpit voice recorder.

"I sure don't want to fly this thing back," First Officer Monreal stated with regard to the return to Chicago O'Hare.

"Well, we're not going anywhere, we're not going anywhere," Captain Bodak assured him and then shut down the engines. "There was like no ice on the airplane that I could really see, though. Could you?"

"I thought it was water. Might have been clear ice, but I don't know . . .," the first officer responded. "I didn't think we were gonna get this SOB right side up."

"When you pushed those props forward that took care of it . . . but we were at eighty-six before, right?" Mike Bodak said, referring to 86 percent of propeller speed required for icing operations, with which they had complied.

". . . I didn't know if you were trying to come back or not or having problems and I grabbed it and I pulled it back, then we were were both . . . I know we were in rhythm because we weren't pulling against each other," First Officer Monreal said, describing how he, too, had grabbed the controls.

"We were just trying to recover, and I kept pushing it forward and I could almost feel the stick pusher going off, too." What the captain might have felt was the first officer on the controls at the same time, because I can find no record of the stick pusher firing during the event, which the first officer confirmed in his next statement.

"It didn't even go off, though it should of."

"I thought I could feel it, and I pushed it down but as soon as I go down I got 'terrain'."

"Did you feel the stick pusher go off?" asked Monreal.

"And that's when I add power. I think I did, yeah," responded the captain.

"They can pull the tape and find out if the stick pusher went off or not," the first officer speculated, referring to the flight data recorder tape.

"All I can figure, you know, is that we got slow in the turn. That's all I can figure."

"You know I looked at that, too. Because we've seen those tapes and listened to them up in training and that was the first thing that I did—I looked and thought I saw 150 on the airspeed," First Officer Monreal said, expressing his concerns with regard to speed in relation to previous incidents, even though the minimum aircraft speed at the time of 153 knots was appropriate, even in icing conditions.

"I've never been so scared in an airplane before! I think we were at a 90, weren't we at a 90?" First Officer Monreal said, referring to the bank angle that would have placed the aircraft on its side with one wing pointing at the ground and the other at the heavens.

"Ah, I don't know. The blue was on the left side, I know that," said Captain Bodak. The "blue" is the sky portion of the attitude instrument, which should be on the top of the instrument with a brown semicircle on the bottom. In a 90-degree bank, the blue would be on the side of the instrument.

The elevation of Central Wisconsin Airport is 1277 feet above mean sea level, so the aircraft was less than 1200 feet above the ground when recovery was finally accomplished after the 20- to 25-second fight. The aircraft was accelerating in its descent before the first officer thought of getting the propeller speed to maximum, which had an effect on the slipstream airflow and possibly helped with flight control in combination with the increasing speed. The pilots discussed the lack of any visible ice on the aircraft before exiting the cockpit to do an exterior inspection.

As he walked around in the subfreezing temperatures, First Officer Monreal still saw no recognizable ice on any portion of the aircraft. He later reported the airplane "was intact and very wet." Captain Bodak was inside the terminal making the proper notifications.

About 20 minutes after Flight 4295 landed, another Simmons aircraft, Flight 4236, arrived at Central Wisconsin Airport after making an intermediate stop in Escanaba, Michigan. The second airliner, a Shorts 360, reported a moderate rime and mixed ice accretion and had been apprised of the poor braking on the ground as the result of the recent passage of an area of freezing rain. The Shorts crew reported no control problems with their aircraft and after parking at the gate the captain did a walkaround to view an accretion of ice. In his report to management made during February 1989, Captain Doug Premuda made the following statement: "Had it not been for your request for this report, I would not have considered the stated flight conditions to be that unusual for a winter operation into an uncontrolled airport."

ATR responded with early conclusions and statements to the airline of no required changes to the aircraft or operating procedures. Simmons Airlines drafted a memorandum to disseminate information on the Central Wisconsin Airport incident on January 23, 1989. Dave Wigand, director of flying for Simmons, made clear the stated position of ATR when he wrote, "Taking into account the information presently available, the aircraft manufacturer considers that nothing needs to be changed on the aircraft or in the operating procedures. This position has the agreement of the French Airworthiness Authority."

For a time in early 1989, the FAA attempted to take a hard look at the developing negative picture of the performance of the ATR-42 in some icing conditions, and this added scrutiny caught the manufacturer in a bad position. The ATR-72 development program was nearing completion, and ATR hoped to have the new aircraft certification in hand by mid-to-late 1989. The mounting evidence of ice-related control problems hanging over the ATR-42 could affect the new aircraft timeline and cost ATR money, particularly if it affected future orders. Something needed to be done to assuage fears quickly.

Operationally and semantically, modifications had already been straining the limits of credulity, and more were still in the works as the Central Wisconsin Airport incident occurred. ATR had determined not to install the larger boots as recommended by numerous experts, insisting they were not needed for the aircraft to comply with current regulations. Discoveries had been made by FAA officials regarding information not shared by ATR that showed some previous statements by the manufacturer to be inaccurate or misleading. Equipment additions and stall-threshold reductions were cutting into the flight envelope, while necessary speed increases in icing were becoming concerns for compliance with other performance requirements for the aircraft. The airplane was trying to tell the regulators something, although they did not appear to be listening.

ATR and DGAC personnel again headed for Seattle, and meetings were held with FAA certification personnel. The French were concerned about the FAA position regarding the operation of the autopilot in icing, because all previous reported incidents and the accident in Italy had occurred while the autopilot flew the aircraft. However, the FAA did not know the exact details of several other incidents that occurred during pilot-controlled flight. For some reason, the FAA refused to act quickly to ban the ATR from flight in icing conditions as it had two years earlier, but it did take the same step it would take after the Roselawn accident to ban the use of the autopilot with ice on the aircraft or when operating in conditions conducive to ice formation. The wording of the 1989 airworthiness directive stated, "Whenever the aircraft exhibits buffet onset, uncommanded roll, or unusual control wheel forces, immediately reduce angle-of-attack and avoid excessive maneuvering." What Captain Bodak and First Officer Monreal could tell regulators was not in the airworthiness directive—that is, when the buffet comes, it is too late.

The latest fix submitted by ATR and quickly approved by DGAC was the installation of "vortex generators" on the upper surface of the ATR-42 wing, just ahead of the ailerons. These credit card-sized pieces of aluminum were part of the ATR-72 design as a result of problems exhibited by the new aircraft in precertification testing. The genesis of the ATR vortex generators had nothing to do with icing tests of the ATR-72 but rather with routine testing that yielded evidence of a disturbing instability of the ATR-72 in some normal turning maneuvers. Vortex generators energize the boundary layer of the airstream by imparting a slight rotational flow perpendicular to the airfoil surface to prevent the separation of the airflow. An aeronautical engineer told me vortex generators were placed on an airplane because the engineers screwed something up in the initial design or the aircraft did not perform satisfactorily in its desired operational envelope.

During the spring of 1989, ATR and DGAC again traveled to Seattle to sell the vortex generator installation to the FAA. They brought another batch of data that showed that there were no problems with the installation and that all tests were satisfactory. Leroy Keith approved the installation, and the rule-making process was begun to mandate their installation on all ATR-42 aircraft operated in the United States. With the addition of the anti-icing advisory system, which involved an electronic sensing device to indicate accretion of ice (a result of the Como investigation), modification of speeds during icing conditions, and inclusion of a minimum maneuvering airspeed, the ATR-42 was ready for another winter.

On April 12, 1989, Robert McCracken of the Transport Airplane Directorate of the FAA in Seattle sent a fax to Roger Anderson in Brussels regarding the installation of the anti-icing advisory system to the ATR-42 indicating, "We are under considerable pressure to issue the NPRM (National Proposed Rule Making) on the anti-icing advisory system as soon as possible." His concern was possibly related to the timing of the ATR-72 certification, and he requested data back from Brussels by June 13, 1989.

The summer of 1989 found the FAA in a scuffle with the French authorities and ATR regarding the ATR-72 certification due to the growing evidence of inadequacies exhibited in icing by the smaller original version. In a certification issue paper for the ATR-72s, the issue was clearly stated:

> *The ATR-72 airplane must be shown to be capable of continued safe flight and landing when operating in any weather conditions for which operation is approved, including icing conditions.*

The FAA cited past events in the ATR-42 and actions taken, concluding in its position,

> *As the FAA cannot issue a new or amended type certificate for an airplane with a known unsafe design feature. ATR must provide data to show that the problems experienced with the ATR-42 will not be present on the ATR-72.*

ATR provided no response to the issue directly to the FAA; however, the DGAC approved the ATR-72 for flight into icing conditions and signed off on the airplane. Without fanfare, the FAA did the same and closed the issue. In other documents relating to ATR certification, the FAA requested specific information regarding the incident history, but there is no evidence this request was ever complied with and the issue was later closed without further comment. It seemed, during my review of the available paperwork from this time, that no FAA officials would go out on a limb like Frank vanLeynseele did with the ATR-42. Questions were raised, but most instances showed only the need for DGAC verification of the product for the FAA to close the issue. The FAA was not going to question a friendly government.

The Air Line Pilots Association, or ALPA, represents more than 41,000 airline pilots flying for nearly 50 airlines. It presented comments during the rule-making process. In letters sent on August 28 and August 29, 1989, ALPA expressed concerns over the handling of ATR icing events and even blasted the proposed fixes as being of the "Band-Aid" variety. The letter of August 28 addressed concerns with airworthiness directive 89-NM-126-AD, which mandated the installation of the anti-icing advisory system. The letter stated:

ALPA accident investigators have reviewed the investigations of one accident and a number of incidents involving loss of control due to ice accretion on the ATR-42 series airplane. Based on our knowledge of these occurrences, we are convinced that the airplane was not properly certified for flight into icing conditions. . . . The aircraft, in our view, is equipped with an anti-icing/de-icing system which is unorthodox, ill-conceived, and inadequately designed. Over the last several years, some cosmetic changes have been made to the aircraft and to the procedures for flight into icing conditions. All of these, in our view, have not solved the problem to overcome the deficiencies of the anti-icing/de-icing system.

ALPA also observed, "All other aircraft types were not certificated for flight in freezing rain as well, yet these same aircraft have not experienced the serious loss of control incidents as the ATR-42 has."

The next day, ALPA addressed the other pending airworthiness directive, 89-NM-101-AD, which mandated the addition of the vortex generators as a means of ending the ban on flight in icing conditions with the autopilot flying the ATR-42. According to the judgment of the FAA, the vortex generators were an "acceptable alternate means of compliance" and provided an "equivalent level of safety" to the pilots feeling what the aircraft was doing by hand-flying in ice. The autopilot can provide relief from the tedium of hand-flying, but its use eliminates the pilots' hands-on feel of how the aircraft is performing and can lead to a dependence that could be dangerous. Most ATR pilots at American Eagle use the autopilot during 90 percent of the flight, and I have seen enough of some of their rusty hand-flying skills to feel uncomfortable with the abilities of certain pilots to deal with the aircraft in an unusual or emergency situation without the autopilot. Even in simulator training, the use of the autopilot is encouraged by the instructors. I have witnessed qualified pilots trying to hand-fly a single-engine instrument approach to minimums and execute a missed approach who put the aircraft into wild gyrations and even into the weeds. A second time, the same pilots used the autopilot with the recommendation of the instructor and the autopilot handled the airplane. The question is, "What happens when the autopilot is not working?"

ALPA supported the intent of the proposed vortex generator airworthiness directive but questioned its effectiveness. Given the intention of the other airworthiness directive of "preventing the ice accretion on the ATR-42 aircraft which in turn would prevent other icing related incidents/accidents from occurring," ALPA mistrusted the reasoning. Its apprehension was indicated in the statement, "We suspect that [this airworthiness directive] is another 'Band-Aid' type fix to an aircraft which was inadequately certified for operation in icing conditions."

The ALPA position letter also indicated the association's concerns regarding what might still be a neglected area—ice on the horizontal tail of the aircraft and how that might have contributed to many incidents and one accident, all of which exhibited control difficulties in the pitch axis as well as roll. While this concern was expressed and suggested action proposed, the FAA did not definitively act on the tail ice question, even years later, after the Roselawn accident.

ALPA posed additional questions to the FAA regarding the airworthiness directive provision that allowed the ATR to operate for a limited time in freezing rain, a ridiculous premise since no aircraft had a certificate to operate in that weather phenomenon. The airworthiness directive did state, "Prolonged operation in freezing rain should be avoided." ALPA's letter queried:

How does the FAA define "prolonged"? In excess of what time frame does the FAA consider "prolonged"? If the FAA cannot define "prolonged," then perhaps it should simply prohibit flight into freezing rain. In past correspondences with the FAA, we have been informed that this aircraft, or any aircraft, was not certificated for operation in freezing rain. Does this statement by the FAA then represent a change in policy regarding aircraft certification for operation in freezing rain? If it does,

what has the FAA done to [e]nsure that all anti-ice/de-ice sys-
tems on aircraft previously certified provide adequate protec-
tion for operation in freezing rain? If this statement was
contained in the AD in error and the FAA still maintains that
this aircraft was not certified for operation in freezing rain,
then the FAA must outline procedures for flight crews to follow
should they inadvertently encounter freezing rain. Since freez-
ing rain cannot be predicted, should pilots refrain from flight
into any icing conditions? What guidelines has the FAA estab-
lished to assist pilots in determining if their aircraft will be sub-
jected to freezing rain? Are there any specific prohibitions
against operations in freezing rain for other aircraft certifi-
cated by the FAA? If there are none, then why is the ATR-42 sin-
gled out?

ALPA was frustrated with the lack of control exhibited by the FAA
and expressed such in its list of questions. In his summary, Harold
Marthinsen predicted future problems by warning, "We feel, however,
that problems will continue to exist on the ATR-42 aircraft when operating in icing conditions" and that the root of these problems was
the inadequate certification of the aircraft for flight in icing conditions.

On August 28, 1989, the day the first of the two ALPA letters was
sent, Jeffrey Gagliano and I were among the group selected at Simmons Airlines to attend ATR first officer training. It was a happy day,
and we were envied by those sent to the Shorts 360 "slave ship." We
had no idea, however, what road we had embarked on.

I don't know whether the FAA ever responded directly to the questions of the pilot union. However, Robert McCracken of the FAA certification office forwarded the ALPA comment letters to the Brussels
office of the FAA with a cover letter that began, "Welcome to the world
of the ATR-42!" Everyone I have discussed this opening sentence with
has the same response: Bob McCracken was saying, "Welcome to my
nightmare!" The ALPA viewpoints were eventually discarded when the
final rule was issued in Seattle based mainly on data supplied by ATR
and DGAC. The rule is dated October 31, 1989, five years to the day
prior to the deaths of 68 people in the crash at Roselawn.

With the incorporation of the two airworthiness directives, the accident in Italy and the Central Wisconsin Airport incident were considered closed. The danger of inattention to the aircraft in ice was
pounded into our consciousness again and again. One instructor at
Simmons told us, "When you get on the line, you will hear the stories
of the evolution of these minimum airspeeds in ice. I want to tell you
they are true. We take them very seriously here. This aircraft will try
and kill you if you get too slow with ice on it."

Icing awareness was a preamble to any speech given by a captain to an unfamiliar first officer. We learned to spend as much time looking at the wing as we did looking at the instruments when we were flying in clouds with ice accreting. "Ice" was a four-letter word to ATR pilots. We were aware. We did our part. The airplane continued to let us down.

On November 1, 1989, the day after the final rule was issued, I reported for my first day of airline flying with an initial operating experience captain. Initial operating experience was the final part of the training process when a new crewmember was allowed to act as a pilot of the aircraft with paying passengers on board. I excitedly reported to the airport two hours before my required time. Captain Paul Nietz found me in the crew lounge operations area and introduced me to our flight attendant for the three-day trip, Sandi Modaff. Later, in the cockpit, Captain Nietz gave me the ice speech for what seemed like the millionth time since I started training for the aircraft back in August.

"If we are in icing conditions, we will bug icing airspeeds. We will not turn the aircraft more than 15 degrees of bank at any speed below yellow bug with flaps at 15 degrees. We will not bank the aircraft at any angle beyond 30 degrees, period. The only time we will be below 170 knots with flaps zero is when we are climbing. If we get below 170 for any reason other than that, we will set 15 degrees of flap. You will not fly this aircraft faster than approach speed plus 10 knots and never below approach speed for any reason except touching down. Is that understood?"

It struck me odd at the time, and in retrospect, I compare the anomalies of the ATR to what I learned about the "Mach tuck" in the Learjet when I was flying for Kohler Company. The Learjet was the first time that I had been told about an aircraft-specific control problem. In the Lear, a number of accidents occurred when airplanes exceeded their maximum operating speed and experienced control problems that led to the loss of control and subsequent crash. This one anomaly was drilled into my head during training to fly the Learjet. One striking difference between the Learjet and ATR problems was evident in my mind. The easiest way to avoid problems in the Learjet is in the hands of the pilot—remain within the operating capabilities of the airplane. With the ATR ice problems, you never knew. What might have been innocuous icing on the last leg might become a problem on the current flight. What you might have seen 100 times before might be fatal, or nearly so, on the 101st exposure. The threshold between controlled flight and uncontrolled flight was unclear. With the Learjet, a red and white

"barber pole" on the airspeed indicator clearly says, "do not trespass." With the ATR, in some types of ice, there is no defined limit.
Many people provided warnings, but no one could explain the
hard envelope or define the threshold of danger. The only issue being pushed was, "Go fast and stay alive." Trainers and managers
did not tell us that tail ice accumulated faster when additional airspeed was used to increase lift for the wing.

In early December 1989, I was paired with Captain Mike Bodak
for a three-day trip, and he told me the story of his incident the previous December. The concern and pain was still obvious in his face.
When he told me the story of seeing his daughter in her holiday outfit as the aircraft was going down, it sent chills down my spine as I
thought of my own three daughters. I could see how the incident had
affected him and was still affecting him a year later. "I have assembled a file this thick," he stated, holding his thumb and forefinger 3
inches apart, "with everything from the BS the company and ATR
gave me to prove their point to my own research. I don't think they
fixed this thing after my incident, but I know I won't be caught again.
I go fast, and I don't mess with the ice in this airplane." I can still picture him sitting across from me in that cockpit as we discussed his
episode. He was a good soldier and kept his mouth shut, knowing
down deep there would be another incident or an accident. When I
next sat across from Mike Bodak, four years later, I could sense his
level of upset at our encounter and the road of challenges he knew
we would travel. It was a road well known to Captain Bodak.

By the fall of 1989 we had new promises for the ATR-42. In addition, the ATR-72 received its certification on schedule, and our first
stretch ATR arrived at Simmons Airlines in August 1991. On June 11
and 12, 1990, a meeting was held in Madrid, Spain, of a committee of
the International Federation of Air Line Pilots Associations, or IFALPA.
One of the items on the agenda was the ATR-42 icing incidents.
IFALPA had issued a warning in an operational advisory letter to its
member associations during November 1989, the intent of which
closely resembled the body of concerns ALPA expressed to the FAA
in qualified response to the proposed airworthiness directives during
August 1989. The June meeting discussed and reviewed the comments of Italian pilot Captain Danilo de Judicibus of ANPAC regarding the ATR anomalies. Captain de Judicibus began his dissertation by
noting the vortex generator installation cannot take care of the worst
icing on the horizontal and vertical tail, which would continue to be
a problem for the ATR aircraft. He theorized, like many other experts,
that the accident in Italy and the first Simmons incident experienced
". . . wild wing rocking [which] was induced by a kind of dolphining,

initiated by the cyclic loss of lift of the stabilizer, due to the anomalous residual icing accretion. . . ." This statement correctly correlated the interaction of all parts of the aircraft and the anomalous effects of structural icing that can result in any number of resultant combinations to produce similar flight characteristics. He restated the conclusion in the context of the growing count of reported ATR problems.

In considering the current accumulated number of icing-related ATR-42 accidents, near-accidents, and incidents, it is mandatory to keep in mind that different effects may be provoked by different actual icing conditions of all surfaces. A wing leading edge effectively de-iced, but with the following 50 to 70 centimeters (19 to 28 inches) of the wing upper part heavily contaminated, in concomitance with a not yet de-iced stabilizer, or a not yet de-iced wing also in concomitance with a stabilizer leading edge clear of ice, but with the two upper and under ridges and the whole of the surface heavily contaminated, is bound to produce similar but not necessarily identical aircraft behaviors.

Captain de Judicibus added, "One of the similarities is the violent lateral instability. Attempting to cure this type of problem by trying to reenergize the ailerons with vortex generators appears to me like providing hooks on which to hang the ice." His remarks close with disconcerting concerns, the last of which hits to the heart of the FAA inaction.

In conclusion, we cannot say that the effects that ice produces on this airplane are identical or conducive to an identical airplane behavior, but for sure, one can say that they spring from the same faulty design concepts. Wing airfoil, and mainly stabilizer airfoil, type and amount of energy devoted to de-icing equipment, amount of booted area, control line design philosophies and construction techniques; these are only a few of the accident-related deficiencies that spring off the Final Report of the Italian Investigative Board for the ATR-42 accident in Como, Italy. It strangely looks like the U.S. FAA is incapable or unwilling to put to good use a serious investigation on this aircraft's sometime strange deportment.

The next set of icing-related incidents for the ATR occurred as a result of improper ground deicing but were significant because they were predicted by the Bascombe-Downs testing in 1988. In testing the tail, the investigators noted the potential for ice and snow clogging an aerodynamic gap between the horizontal stabilizer and the elevator. They recommended a shield for the gap to prevent the possibility of this critical area becoming contaminated. ATR apparently did not give credence to the report and ignored the suggestion.

During January 1991, Simmons Flight 4160 was scheduled to depart Madison, Wisconsin, bound for Chicago. Captain Al Bennett and First Officer John Walters arrived at Truax Field to retrieve their aircraft from the heated hangar where it had been stored overnight. The warm airliner did not require deicing, and John Walters did a thorough walkaround inspection before boarding the passengers and taxiing out for departure. The pilots exercised the controls just before taking the runway, ensuring complete freedom of travel, and reported no binding or unusual feel. The aircraft accelerated down the runway, and everything progressed normally until shortly after the crew rotated. On the climb, John Walters reported something feeling strange in the controls and he turned the aircraft over to Captain Bennett. A strong control pressure seemed to worsen as the aircraft accelerated until both pilots' strength on the control column was needed. The flight advised air traffic control of their problem and wrestled the aircraft around for an approach and landing. Inspection on the ground revealed an accumulation of water on the horizontal tail that had frozen when the aircraft was moved from the warm hangar to the cold outside.

On October 6, 1991, a similar event occurred to Flight 4237 on departure from the maintenance base in Marquette, Michigan, when Captain Mike Uren reported a dramatic pitch up of the aircraft after rotation that required both pilots' strength on the controls. Once again an emergency landing was made after several minutes of frightening control problems. The crew was uncertain what was happening. The faster they went, the higher the control force. On inspection of the tail from the deice truck, maintenance personnel discovered snow and ice in the aerodynamic gap. This was the same aircraft, N429MQ, involved in the Madison incident. In a report of the chronology of icing incidents prepared years later for company and FAA personnel, Glenn Leonard, AMR Eagle ATR fleet manager, wrote, "Two occurrences at Simmons with contamination in elevator control gaps due to improper ground deicing procedures. These incidents posed no serious threat to the immediate safety of flight and were not unique to the aircraft design or actual flight procedures." While the last portion of the sentence is almost nonsense, the classification of the events as not posing any serious threat to the immediate safety of flight is beyond reason.

On October 18, 1991, Jim Menard of Simmons Airlines' field services department issued a warning to station managers and ramp personnel. "Effective immediately, an added step is necessary when deicing the tail of every ATR. This change is prompted by a number of incidents reported by the manufacturer (as well as one here at MQ) wherein the ATR's tail has been found to be more sensitive to ice than was originally believed."

Operation of the ATR requires the vigilance of the deicing crew. A conscientious crewmember visually inspects the underside of the tail to ensure the deice fluid flows through the entire gap as required by the service bulletin. However, since the events of 1991 at American Eagle, vigilance waned early on. Deicing is a dirty, cold job and getting glycol all over your face and body makes you a sticky mess. So who can blame the operator for a "long-distance inspection?" The pilots, flight attendants, and passengers of the next incompletely deiced ATR, that's who.

In the spring of 1996, I was informed of a Trans States ATR, operating as TransWorld Express, that was forced to make an emergency landing "short of St. Louis" for this apparently ongoing problem. According to the Trans States pilot source, the crew reported that the aircraft was "not de-iced properly," had "full trim input and [they] still could not do much with [the aircraft]."

The operation of the ATR is limited to a maximum altitude of 25,000 feet. For the routes the aircraft operates, this is normally sufficient. Rarely do we see "flight levels," those altitudes above 18,000 feet, and when we do, it is usually a short-lived apex of what I often referred to as "the flight path of a golf ball." With the exception of vertically developed clouds such as thunderstorms, which carry a large quantity of moisture from low to high levels of the atmosphere, the clouds above 18,000 feet are usually layered and low in moisture content. Temperatures at that altitude are normally below zero and too cold to allow significant icing.

The average temperature on the ground during August near Wales in the United Kingdom is 60 degrees Fahrenheit (15.5 degrees Celsius). With normal rates of cooling with altitude, referred to as lapse rates, the anticipated temperature in an average atmosphere at 18,000 feet over Wales in August is minus 6 degrees Fahrenheit (minus 21 degrees Celsius). Normally, an icing-related incident is unanticipated in those conditions. As a Ryanair crew found out, however, and as we as pilots understand, the realm of flight is rarely, if ever, standard or highly predictable.

On August 11, 1991, an ATR-42 operated by Ryanair was in cruise flight at 18,000 feet with an indicated airspeed of 180 knots and the autopilot on. As the aircraft progressed toward its destination of Stansted-Waterford, the flight entered a thin, altostratus cloud layer over the Brecon Beacon mountain range. The pilots were aware of forecast moderate icing in clouds and had the ATR's level 2 anti-ice system operating, with the exception of the propeller speed, which was maintained at normal cruise of 77 percent.

The sun was visible through the thin clouds, and the temperature was at 14 degrees Fahrenheit, still below freezing, although about 20 degrees warmer than average. Recorded cockpit conversation indicated the crew realized the accretion of ice had begun, and they reacted properly by activating level 3 deicing. Two minutes and 30 seconds after the activation of the deicing boots, the aircraft decelerated from its cruise speed of 180 knots to 145 knots with no input from the crew. Power remained constant. At 145 knots, the autopilot disengaged, and the aircraft departed level flight with a developing roll instability. The pilots reacted rapidly with attempted control inputs, including a pitch up that resulted in a rapid roll to 50 degrees of bank. The crew lowered the nose, and airspeed increased as the airliner lost about 500 feet of altitude, wings rocking until the airspeed built back to 180 knots, when controlled flight resumed. The stick shaker warned the crew of an imminent stall, but it is unclear whether the stick pusher intervened, even when the angle of attack reached 18.5 degrees, far beyond the threshold for pusher activation.

To rid the aircraft of the residual ice that had built on unprotected surfaces of the wing, the flight descended into warmer air and continued the remainder of the flight without incident. This crew had violated operating procedures that called for a minimum 86 percent of propeller speed in icing conditions as a part of level 2 anti-ice protection. The disconcerting factor goes to the variety of circumstances that lead to the same control excursions, exactly as predicted and cataloged by Captain de Judicibus, more than one year earlier.

Ironically, on April 17, 1991, Air Mauritius was operating an ATR-42 at 16,000 feet over the Indian Ocean when a similar encounter caused the aircraft to roll up to 40 degrees. Although the same circumstances were present, the airspeed loss was more rapid and the loss of control was at higher airspeed. The aircraft decelerated from 183 knots to 160 knots in two minutes when the autopilot began sensing the lift anomaly develop, and digital flight data recorder traces show the beginnings of aileron deflections and minor roll oscillations just before automatic disengagement of the autopilot. Once again, the onset of the control problems were rapid and masked by the autopilot. In both cases, it is believed the vortex generators were ineffective in preventing loss of control or assisting in the recovery from loss of control.

Basic airmanship requires the pilot to monitor and correlate the proper indicated airspeed and amount of power to maintain that speed at various altitudes and temperatures. Good pilots react to those changes promptly. It is impossible for pilots to predict when the ATR is going to have a control anomaly, and they have to expect the unexpected whenever operating in ice.

In the fall of 1992, Captain Bob Augustson and First Officer Bob Crawford piloted a Simmons Airlines ATR that was approaching the

Michiana Regional Airport outside South Bend, Indiana, after a 60-mile flight from Chicago O'Hare. The aircraft was accumulating ice and was configured for landing when a shuddering vibration began from the tail of the craft. Captain Augustson advanced the power levers and ordered a reduction of flaps. As the aircraft accelerated and the flaps retracted, the shuddering lessened. Augustson had thrown out the approved flight manual, correctly identified the onset of an ice-contaminated tail stall, and responded appropriately, if unconventionally. Simmons ATR pilots are only trained to recognize and recover from a wing stall, yet the horizontal tail also is a flying airfoil and subject to the same physical principles of lift as is the wing. Captain Augustson did not produce a wing stall when he raised the flaps and increased the wing angle of attack. This same maneuver, that is, raising the flaps and causing an increase in the angle of attack of the wing, is what investigators conclude precipitated the Roselawn accident. Bob Augustson did not report this event to the company or FAA.

On March 4, 1993, at 5:04 P.M. Eastern Standard Time (EST), Continental Express Airlines Flight 3444 was en route between Baltimore-Washington International Airport and Newark, New Jersey. The flight was being vectored for an approach to runway 4R at Newark, and the autopilot was handling the airplane for the first officer. The ATR-42 carried 31 passengers and a crew of three and was operated by Britt Airways under a licensing agreement with Continental Airlines. Captain Brian Gignac was new to the ATR, with only 300 hours, but he had more than 9000 hours of flight time. First Officer Thomas Larnad had fewer hours in the ATR and 5700 hours during his 21-year career. The crew activated all anti-icing and deicing equipment during their descent and began accreting ice. The captain later reported turbulence as moderate in the clouds with rime and mixed ice accumulating at a light to occasionally moderate rate. Weather at the airport was reported as a broken cloud layer at 1600 feet, overcast layer at 2700 feet, and a visibility of 1.5 miles in light rain and fog. The ground temperature was 38 degrees Fahrenheit with winds out of the east at 80 degrees at 26 knots with gusts up to 45 knots. Pilots had reported icing and turbulence throughout the area, and the crew was anticipating windshear and landing at an icing speed of 129 knots. Cumulus cloud shafts punctuated the widespread stratus cloud deck that occasionally broke to allow the pilots a glimpse of the New Jersey countryside.

The flight was cleared for the approach at a maximum speed of 180 knots. Just before the final approach fix, about 7 miles from the airport, First Officer Larnad reduced power to begin slowing the aircraft for the upcoming configuration changes of flap and landing-gear extension. Simultaneous with the speed reduction, the autopilot reacted to airflow anomalies by correcting for various bank excursions with aileron inputs that were becoming familiar on ATR digital flight data recorder readouts. The angle of attack began increasing slowly, the autopilot disengaged at 168 knots, the aircraft entered a rapid right bank reaching 52 degrees, and the nose headed earthward. Both pilots grabbed for the control column and used full left aileron along with a rudder deflection to right the craft. As the airspeed built above 170 knots, the right wing began to rise to level flight, and the first officer continued flying the aircraft. Within seconds, the aircraft executed another right-wing-down bank, and recovery was accomplished a second time. Both pilots searched the overhead panel and console to verify trim position or an indication of a deployed spoiler. Nothing was out of the ordinary, and a few moments later, the aircraft departed controlled flight for a third time, was recovered, and flown to landing by the captain. A fourth unexpected bank was less dramatic, and the captain reported that the controls became spongy and that an airflow disturbance over the ailerons could be detected in the control column just before the roll. With each event, the aircraft lost about 100 feet of altitude.

The flight descended below the cloud deck, and the pilots extended the flaps and gear and sighted the airport. There were no more uncommanded departures from controlled flight. Crew observations after the onset of the control anomaly revealed a previously unseen ridge of ice that extended beyond the boot as far back as the crew could observe on the top of the wing. The accumulation, according to Captain Gignac, had formed almost back to the "fat part of the wing," and the crew reported ice had formed on the front side window of the cockpit. This incident report is the first to officially mention this type of ice formation on the side window.

Two years later, this ice was called "trapezoidal ice formation," which indicated freezing rain or freezing drizzle exposure. ATR claimed this discovery was identified for the first time in 1995, yet the report of the Newark incident written by ATR states, "Crew reports—cockpit side windows obscured by a thick layer of ice." Both pilots reported freezing drizzle or freezing rain, but the first officer reported no visible accretion at 3000 feet on the approach with outside temperature between 32 and 34 degrees Fahrenheit.

ATR Support, Inc., was a party to the investigation, and ATR ruled that the encounter was caused by severe turbulence, an evaluation the FAA and NTSB accepted. Despite the mounting evidence, the FAA, which is responsible for monitoring incidents and ferreting out trends, did not take notice, and the stage was set for tragedy. All of the indicators were unmistakable, but nobody questioned the findings and judgment of the manufacturer. The key to this event was ATR's claim of severe turbulence from what they saw on the data recorder when both pilots reported only moderate turbulence. Even though there is no quantitative way of determining the severity of turbulence, or icing for that matter, the personnel of the aircraft manufacturer more than 4000 miles away deemed themselves better able to judge the degree of turbulence than the competent airline pilots who rode through it. The modification of the facts was not new for ATR; it is not known whether ATR requested an aircraft structural inspection after an encounter with severe turbulence as required by the *Aircraft Maintenance Manual.*

Another incident report began:

There we were, cruising at 16,000 feet with our anti-icing and deicing equipment selected "on." Neither the captain nor myself knew that, in this specific aircraft, with anti-icing selected, the trim whooler would periodically sound. I deselected the autopilot, and the whooler continued to sound without any trim inputs from me. We felt this was the early stages of a trim runaway. The captain gave me the checklist as I selected the autopilot on again. He suggested I locate the trim runaway checklist and the trim motor circuit breaker. He took control of the aircraft with the autopilot engaged. As he took the aircraft, he noticed a strong shaking with the control wheel, so he pressed the button that temporarily disconnects the autopilot. As soon as he did this, the aircraft rolled violently to the right to about 45 degrees, and just as violently to the left to 45 degrees. The captain's control inputs were full opposite aileron and rudder. As he did this, he disconnected the autopilot. It made a couple of more violent banks in each direction as we began a descent. We notified ATC of our problem, requested an immediate descent and the current weather in Quincy. As the controller gave us the weather, the banks ended and we regained full control of the aircraft. We notified the flight attendant and told her to sit down. We then began to troubleshoot the problem. Our first thought was some sort of icing. . . .

This NASA report, written by the first officer of a Trans States Airlines ATR, operating as Trans World Express, described events that occurred on September 14, 1993, and appeared on the NTSB list of incidents as both an "icing episode" and "related event" to the Roselawn accident.

On November 24, 1993, the Certificate Management Office of the FAA, which was located in Dallas and was responsible for the super-

vision of Simmons Airlines, began an investigation into the icing incident Dan Rodts and I had experienced in Marquette on November 23.
Contact was made with Bob Ballaster in the Chicago office of the
NTSB; Ballaster was informed of the occurrence and the fact the Certificate Management Office believed the incident to be significant.
Ballaster stated the NTSB was not interested in the report nor would
the NTSB become involved in the incident. Believing a serious situation was developing as a result of ice problems, the Certificate Management Office continued to pursue the investigation and contacted
Jim McDonald of the Aircraft Evaluation Group of the FAA in Seattle
on November 29, 1993. McDonald showed interest in the matter and
requested that his office be kept advised of the ongoing investigation.
His staff was requested to attend a meeting to be held on January 20,
1994, in Chicago to study the incident further. I was told of the meeting several days before it was to be held and drove to Chicago on the
morning of the meeting.

When the meeting was called to order, 15 people were present.
The FAA was represented by Jim Anderson, Jerry Barron, and Russell
Nance, all from the FAA Certificate Management Office in Dallas.
AMR sent Ed Harvey, Simmons' director of flight operations, and Al
Rumph and Glenn Leonard, both ATR fleet managers, Al for Simmons
and Glenn for AMR Eagle. Dan Rodts, Sue Larson, Julie Bonk, and I,
as well as Captain Duane Hastrich, First Officer Chris Nielsen, and
Captain Mike Bodak were there. The ALPA sent Tom DeMaeyer and
Jeff Sedin. Invited parties from the Seattle office of the FAA and rep-
resentatives from ATR were absent. Jerry Barron stated the Aircraft
Evaluations Group (AEG) was going to attend, and he did not know
why they failed to show.

Al Rumph indicated ATR was in its fourth revision of what happened to our flight, and they did not want to discuss the matter before
finishing the report. I nudged Dan and said loud enough for others to
hear, "Fourth revision? What the hell is that?" I received a stare from Ed
Harvey. Shortly thereafter, I noticed another look from Mike Bodak. It
was a look of frustration and disgust. In front of him lay a tattered
manila file folder containing perhaps 200 pages of documents.

During the meeting, I was the central focus of questioning, because I had been flying the aircraft and had actually felt the instanta-
neous onset, prolonged duration, and eventual conclusion of the
control oscillations. The FAA panel questioned me thoroughly, and I
was surprised nobody from the airline or ALPA had any inquiries.
Dan, Sue, and Julie all answered questions, as did the Flight 4227
crew. When it came time to seek information from Mike Bodak, Ed
Harvey derailed the questions. The discussion of the Central Wiscon-

sin Airport incident was cut short when Mike began going into how he had sought further expert opinion after his incident and its subsequent closure and public resolution.

I described the buffet frequency as roughly double that of a wing stall and Ed Harvey piped in, "Have you ever stalled an ATR?" I replied "No," but I told him strongly that I had stalled other turboprops and even a couple of jets to the onset of the aerodynamic buffet, and I was estimating the frequency. I also described it at about half the frequency of the stick shaker. It was easily definable to those who had experienced both types of events. During a break, I spoke with Jim Anderson of the FAA, and he said, "I truly believe you had the onset of an ice-contaminated tail stall and that you guys are lucky to be here today. Most encounters with that type of control problem don't have a happy ending. Good thing you were hand-flying the airplane. You got the warning in time."

Jim Anderson was right. The masking effect of the autopilot could have caused us to miss the initial buffet until the aircraft suddenly and violently pitched nose down when the autopilot disengaged. The story was told in the lead of an article titled "Watch Your Tail" in the December 1993 edition of the *Business & Commercial Aviation* magazine. In the article, a picture submitted by icing expert Porter Perkins from NASA showed an accumulation of double-horn mixed icing on an unprotected airfoil.

> *Consider this hypothetical event: An airplane approaches for landing in instrument conditions. The outside air temperature during the approach is 30 degrees F, and accordingly, all of the anti-ice and deice equipment is on and functioning. Thirty-knot wind gusts have been reported on final, so the crew has decided to increase airspeed 15 knots for the approach. All systems are operating normally, and there is no indication of a problem incubating. Five miles from the outer marker, the first increment of flaps is lowered, and landing gear follows. The final approach checklist is completed, and the airplane is in stable flight, on altitude, on speed, and on course. A half inch of ice is encrusted on the windshield wiper arms, but the wing leading edges have been kept clean. Icing conditions appear to be light or possibly moderate. As the glideslope pointer descends, the pilot flying (PF) calls for final flaps, and the pilot not flying (PNF) complies by moving the handle to the full down position. The flap indicator registers the flap movement, and as the needle passes the half flap tick, the control column suddenly jumps forward as if pushed by an invisible hand, pitching the airplane down abruptly. In*

*less than three seconds, the airplane assumes a near-vertical
pitch attitude as the startled pilots wrestle against the control
column. They have no idea what caused this sudden emergency and if questioned later, would still not know.*

That magazine appeared in my crew lounge mailbox after our
event, with a note from Dan Rodts indicating it was from him. As I
read the article, my blood ran cold. I read a section of the article, under the heading "Buffet Warning," with great intensity.

*When a tailplane stall occurs, it may telegraph its presence
with identifiable symptoms. The most likely symptom will be a
buffet through the airframe or the yoke—or both. Some have
reported that the tailplane stall buffet is at a higher frequency,
perhaps double that of buffet associated with wing stall. Certainly, any vibration or buffet from the tail must be taken seriously if the airplane is in icing conditions during approach.*

Three factors were listed that increase the susceptibility of the
tailplane to icing-related anomalies. These factors are aerodynamically
balanced or assisted elevators, tailplanes with small-radius leading
edges, and highly effective wing flaps. The ATR has aerodynamically
assisted elevators (and ailerons), a small tailplane leading-edge radius,
and highly effective wing flaps.

Our January 20th meeting closed with the FAA representatives
convinced of the need to continue the investigation; however, the
lack of support provided by the NTSB and the AEG left the Certificate
Management Office with limited options. Unknown to me, another
meeting was scheduled during the second week of February 1994 to
further discuss the event, which was still considered an icing incident.
The meeting was convened at the Simmons Airlines headquarters at
the Dallas-Fort Worth Airport. I was not invited. It is my contention
that the decision of Simmons Airlines to not request my attendance
was in part the result of my outspoken position at the previous meeting and insistence that training be added to the curriculum for the
identification and recovery from the tailplane stall situation. The manufacturer did attend this meeting. An infuriated Dan Rodts filled me
in on the details a few days later when I ran into him in the operations area of the crew lounge.

"Dan, any word on the investigation?" I queried.

"Steve, it is such BS. We just had another meeting in Dallas and
ATR came and gave some bogus explanation about heavy turbulence.
I thought you would be there, too. When I got there and didn't see
you, I said, 'Hey, where's my first officer?' and they said, 'We decided
not to invite him,' and I knew what was coming because you had
been so opinionated in Chicago."

"But there was no turbulence," I said, looking around the room and noticing the attention focused on our conversation by other pilots.

"I know. I protested this vehemently, but nobody was listening. It's case closed. Heavy turbulence."

"But there was NO turbulence," I asserted. "What does ALPA say about this?"

"I talked with Tom DeMaeyer [the ALPA safety committee representative] and said, 'Tom, what is this BS?' and he comes back and answers with another question saying, 'What do you want us to do, shut down the airline and put 600 pilots on the street? Don't push it,' and I was so ticked off I just walked away."

"What a bunch of garbage. Another coverup and we're just supposed to take this?" It was a statement more than a question. I knew there was nothing left to be done.

The notes that FAA representatives took at the gathering indicated their frustration and stated the meeting closed "without any affirmative action being outlined or completed." The synopsis also concluded, "The Dallas-Fort Worth Certificate Management Office continues to be concerned with the potential icing problems on the ATR aircraft."

Jerry Barron later stated, "ATR pooh-poohed it all, classified the Marquette incident as a 'non-event' and even recommended drug testing for the 'jittery' crew. ATR was not receptive to the idea that something was wrong with their aircraft."

Our case was closed. Turbulence.

On January 28, 1994, a Continental Express flight was in cruise flight at 16,000 feet with all deicing and anti-icing systems on and operating normally. This incident was remarkably similar to the Ryanair event of August 11, 1991. The flight decelerated from 200 knots to 145 knots over a period of nine minutes while flying on autopilot. The autopilot disengaged at 145 knots, and the ailerons deflected on their own to roll the aircraft. The crew recovered from the loss of control, and even though all operational procedures were proper, ATR blamed the crew for lack of proper awareness and operating below the minimum speed for icing conditions. Considering the aircraft was in level flight, the minimum speed for straight and level was not a potential problem at 145 knots. This little-known incident was the last warning that went unheeded before October 31, 1994, when Eagle Flight 4184 rolled in icing conditions and plummeted 9300 feet in 33 seconds.

The path leading to the Roselawn, Indiana, crash had numerous forewarnings and had the FAA intensively investigated just one of the precursors of Flight 4184, the crash might not have occurred.

ATR issued a report for the NTSB hearings on Flight 4184 that listed the accomplishments and milestones of both models of ATRs.

At the end of 1994, 268 ATR-42 aircraft were operated by 53 airlines, and 136 ATR-72 aircraft were operated by 29 airlines. These 404 aircraft were operated in 37 countries and had amassed 3,290,300 flight hours during 3,909,500 flights. An exhibit in the report titled, "Icing Conditions Experience," listed operators that ATR reported as having no "relevant incidents." The carriers were Karair of Finland, CityFlyer Express of Great Britain, Inter Canadian of Canada, Lot Polish Airline of Poland, Cimber Air of Denmark, and Eurowings of Germany. The cycles operated by these six carriers totaled 711,000 flights, roughly 18 percent of the total. If this list is accurate, every U.S. operator has reported relevant incidents. Of 53 airlines that operate the ATR-42 and 29 airlines that operate the ATR-72, some of which operate both, having only six airlines report no relevant incidents is not a positive statistic.

Note that a relevant incident was not defined in any of the documentation I reviewed. Would a relevant incident be a roll event like Flight 4184 or any icing-related control anomaly that could not be blamed on pilots, severe turbulence, or severe ice? It is anybody's guess.

Most passengers on incident flights never realized how close they came to being a national headline. In his account of the 1988 Central Wisconsin Airport incident, which included violent and unusual aircraft gyrations, Captain Mike Bodak added a remark stating, "No passengers or crew were injured, and we were told only one of our 34 plus passengers came to the ticket counter to inquire what happened." After our aborted flight to Marquette in November 1993, the five passengers who were unexpectedly forced to spend a night in Green Bay never noticed anything unusual in the cabin. Even our two veteran flight attendants failed to suspect the extent of our problem.

I am disconcerted by the stonewalling of the FAA and the appearance of seemingly innocuous statements or handwritten notes in the margins of released documents, which, in aggregate, are damning. Obtaining information from the FAA with the use of the Freedom of Information Act is more difficult than getting information from any other governmental agency. Experienced media personnel tell me that the FAA lived up to its reputation in this case. The agency engages in marginal compliance with the law and often distributes documents among various requests to attempt to create a discontinuous story, as they did with those associated with the ATR. I would travel to a source to obtain 1000 new pages of documents, assume I had seen the entire file, only to visit another source who requested the same documents from the FAA and received a different 1000 pages of information. When combined and sorted into an organized chronology, the documents clearly summarized a story of FAA inaction in the face of incontrovertible evidence of the ATR's continuing susceptibility to icing, which became an even larger part of the tragedy and waste of American Eagle Flight 4184 and places in jeopardy every single passenger, pilot, and flight attendant who continues to fly on an ATR in the ice.

10

Mental torture

The CBS news magazine show *48 Hours* aired a story on Thursday, March 2, 1995, about Lauren Anderson, age 18, a survivor of American Eagle Flight 3379, which crashed near Raleigh-Durham, North Carolina, on December 13, 1994. Ten minutes into the program, the broadcast showed two smiling American Airlines employees walking into the Anderson home carrying an intact cloth suitcase and delivering the bag to Lauren. In front of the cameras, the college student opened her undamaged luggage and displayed the contents, each piece of clothing cleaned and wrapped in clear plastic bags. She even handed a paper from one of her courses to her father, proudly proclaiming, "Here Dad, that's the paper I got an A on." Lauren's story was unique because she survived.

The smiling employees who brought Lauren's possessions back to her that day were members of American Airlines CARE team, volunteers who are assigned to families involved in an aircraft accident. Roger Hughes normally works as a mechanic, and Darryl Broderick is an international sales supervisor. Their CARE team assignment was to provide family members with emotional support, compassion, and anything else they might need.

Lauren's parents also were smiling that day. The CARE team had provided them with airline tickets to visit Lauren in the hospital, made motel and automobile arrangements, picked up the tab, and allowed them the time to be with their recovering daughter. Things were all smiles at the Anderson home, and the CARE team had a success to display to the media.

The American Airlines CARE team was the company's response to its observance of the positive aspects of the way Delta Airlines handled the aftermath of the crash of Flight 191 in Dallas in 1985, according to Anne McNamara, senior vice president and senior counsel at AMR Corporation. She related that during the aftermath of Flight 191, the airline noticed it was the first time the issues of the family members were focused on. The CARE team became part of American's disaster plan that goes into effect in case of an aircraft incident or accident. Team members are mostly volunteers, because the CARE program maintains only a skeletal structure and minimum staff during "peace time," as McNamara described it. American has a training program for its employees who want to participate, and they may come from any area of the airline.

Outside the Anderson home, American Eagle spokesperson Marty Heires also spoke on camera, defending the team's efforts against allegations by less fortunate family members regarding the flow of personal information to the airline and insurance company that later could be used against them at time of financial settlement. Heires stated those charges were baseless because there was no contact between the CARE team and the lawyers, which may have been true, although the day the team brought the luggage back to Lauren, they were scheduled to be accompanied by Robert Alpert, the airline's insurance company representative. When he heard that *48 Hours* was at the Anderson home, he decided not to come with the team and showed up later, after the camera crew had left.

The family of Scott Johnson was not so pleased with the work of the American Airlines CARE team or the actions of the airline as a whole. Six weeks after the Flight 3379 accident and without warning, Katie Hales received some of Scott's belongings back from the airline. She found his wallet with credit cards, a roll of Lifesavers, Scott's airline ticket, and his keys "rattling in a Federal Express box." At the time of the *48 Hours* taping, she was fighting with American and its insurance carrier over the return of the rest of Scott's personal items contained in his suitcase and briefcase. The airline said it had not received any of Scott Johnson's clothing and that it "does not return 'significantly damaged items' because it does not want to cause any more trauma to the families."

Lauren Anderson appears to have been a CARE team success, including the near-miraculous return of her luggage, but the nightmare of Scott Johnson's family, even with the return of a limited number of personal effects, more closely resembled the experiences of the families of American Eagle Flight 4184. The balanced news report, in toto, also gave the impression there could be potentially questionable motives for placing airline employees in families' homes during such a vulnerable time for the grieving relatives.

Officially, no tie exists between the airline's insurance and legal advocates and the CARE volunteers, but many families, including the Johnsons, have made claims regarding the information about the family that shows up during settlement or legal discussions. It is also possible the airline employs private investigators to look into the family's affairs and history. In any event, the airline will probably never admit to hiring private investigators, and the families will never know.

American Airlines' emergency plan is carefully prepared, updated with new information whenever it becomes available, and widely distributed. However, it is not something practiced often. Many times, even an innocent act of an employee with good intentions blows up in the eyes of the media or the families. Unlike the training the pilots and flight attendants receive on a routine basis to handle inflight emergencies, no all-encompassing plan exists for the diverse and fluid emergencies of an accident.

American Airlines has aircraft in the air 24 hours per day. The company's exposure is fantastic, and it does not know whether the next accident is going to happen to one of its airplanes or that of a competitor. If it does happen to an American Airlines flight, it could be anywhere on the globe, in just about any culture or climate. The airline does not take this responsibility lightly.

People care at the highest levels of American, giving the airline a corporate soul, but American Airlines exists for one purpose, to make a profit for its stockholders. Transportation companies do not have the right to exist, nor do they have a responsibility to the public as a reason for their existence like fire or police departments. Besides, in high-stakes business, little room is available for the heart over profit. A company with nearly 100,000 employees in various subsidiaries around the world cannot control the actions of every single one, no matter how hard it tries.

Nearly all of the families of the Flight 4184 crash initially told me that they felt nothing could be as hard to take as the initial confirmation of their loss. Most believed their original assumptions proved wrong.

On the morning after the accident, two CARE team members showed up at the Anglemyer home in Rocky Mount, North Carolina, and introduced themselves to Sandy, who will never forget the young man and young woman who entered her living room. While the man was dressed in a suit appropriate for the visit to a recent widow, Sandy said that his partner, a female employee of American Eagle, was inappropriately dressed in a hot-pink short skirt. According to Sandy, she prattled on incessantly about their lack of training and lack of understanding about their role. "We don't know what we are doing, but we're here," Sandy remembered the woman saying. Even in her emotional distress, Sandy quickly determined she could not deal with these two individuals and left them to her family and a friend.

Not every family was greeted by an immature and unprepared volunteer as Sandy Anglemyer had been, but some reported a similar discomfort with the type of questions asked by the volunteers. Many people don't know what to say when they talk to someone who has experienced a great loss, but if these volunteers are indeed trained and not a part of the airline loss-control team, why would they be so callous to ask personal questions about the health of the deceased and the status of family relationships? One family member was distraught about the CARE team members going into the victim's bedroom to look around, while another told of an almost ceaseless question-and-answer period regarding the photographs around the home.

Not everyone reported a negative experience with the CARE team. Numerous families formed an emotional bond with the volunteers. Many pitched in to prepare meals, help with household chores, arrange travel, and see to the numerous other needs of the family. The volunteers, however, are not trained psychologists or grief counselors, and many could not provide what relatives needed most—straightforward answers.

Many families wanted to visit the crash site, not knowing the level of destruction that existed there. Some thought they could aid in the identification process and help return their loved ones home for burial. Others felt the accident would not be real until they saw the location. Five members of the Grimberg family traveled from Colombia with arrangements made by American Airlines and pulled up to the Indiana State Police roadblock in a van supplied by the airline. They were turned away before getting close to the field. The group drove to the National Guard Armory in Remington to identify the remains but were once again turned away. The airline had worked with the family to get them close, but the grieving group returned to South America without seeing anything of Semmy, Guilda, or Adrian.

The family of Wan Suk Ko came from Korea and rode to the same roadblock in a van with the family of Frank Sheridan, who drove from their homes in Illinois against the advice of their CARE team members. Pat Hansen, Frank's sister, had said, "I don't care what you think, I need to go." Ironically, Frank Sheridan and Wan Suk Ko sat next to each other during the flight.

The CARE team volunteers told the relatives to ask them any questions and told the families they could provide answers that were more accurate than those given in press reports. The airline had access to all parts of the investigation and knew facts that were not made public. However, some families grew suspicious of the airline's motives.

The airline did have its interests to protect. Reports with credible information were rare, and some reports inaccurate. One example centered around the condition of the remains. The television showed the destruction, but many families felt that the scene must have been photographed after the bodies had been removed because none were visible; in reality, the bodies were still there. Relatives reported being told by their CARE team member that there were intact bodies, and they would get their loved one back, only to learn the painful facts later.

When questions were asked, the CARE team volunteer called the company for information, and often days passed before the answer came back. Some families had extensive contact with the team, and others only had a visit days after the accident. Families gathered to plan a funeral and then were told it would be weeks before they received any remains, so they could not plan services. Many people who traveled to Indiana went back home and could not return for any memorial services. Funerals are for the living, and to accomplish closure with any death, especially unexpected death, families need the healing provided by the ceremony. These people were deprived that closure in many cases because the crash was so destructive.

Once families learned the level of destruction, questions were raised about the disposition of the crash site. Friends and relatives understood the site was a farm and sought answers from the volunteers. The airline assured families, although this point is strongly contested on both sides, that American was in the process of buying the field and that the land would be set aside and never farmed again. When the families later learned this was not the case, many were irate and distressed over the fact the farmer intended to plow and plant land that was, in essence, a burial ground.

As it turned out, even the timeliness of the information was not what was originally promised. Family members saw news items and questioned CARE team members about them. The CARE team queried the office and then admitted the reports were correct. The team had assured relatives they would get the word on any developments first, but by the time the "board," as they called it, cleared and culled information, it was often too late to preempt the media. In addition, any answers they did receive often seemed to be carefully crafted.

Finally, near the middle of November, the airline tried to arrange for family members to see a "grief counselor" to discuss issues. Barbara Tribble's son Garvin, a new attorney, recognized the inherent conflict and advised family members of a simple legal fact. The relationship between the counselor and the families was not covered by the same ethical bonds that exist between a doctor and patient, and anything said in the sessions could be, and most likely would be, subject to discovery by the airline when lawsuits moved through the discovery process. The entire situation sent warning signals to the new attorney; however, the Tribble family decided to attend the session.

The counselor excused himself during the session and returned to state that the remains of the victims were being released that day. The family had just sent out-of-town relatives home because the previous day they had been advised it would be more than a week before the remains of Barbara Tribble would be coming home.

Al and Cindy Gagliano were warned by ALPA to move any of Jeff's technical items out of the house, because the airline might move to seize the manuals, which could reveal damaging operating information later on. This revelation added to the distress of the family, because the CARE team volunteers had just returned from a trip to the basement to see Jeff's room.

A disturbing event occurred during the funeral services for one of the flight attendants, which was attended by the family's CARE team. The team members brought along their program supervisor. When the flight attendant's sister was introduced to the supervisor, he said, "Yeah, I know who you are. And I know who your husband is because we have files on all of you." The conversation came to an abrupt halt.

Finally, during a memorial ceremony in August 1995 to dedicate the mass grave memorial, Ron Tribble was approached by his CARE team member, who asked, "By the way, how are your plans coming for moving to Colorado?" Ron and Barbara had planned to move there before she died.

Overall, the response to the CARE team was positive on a personal level. The families felt genuine caring from some of the volunteers, which seems obvious, because individuals would not offer their services in such an emotionally destructive situation if they were not caring and compassionate. The problem came over the lack of information that was promised, and after a while, most of the families understood it was time to end the relationship with the CARE team. The airline also has a policy of terminating the relationship when the health of the volunteer may be affected.

When the time came, families reportedly were told the volunteers could no longer communicate with them. Although some relatives said "Good riddance," others who had become emotionally dependent on the relationship with the CARE team member felt another loss. Perhaps professionals with no ties to the airline would be a better alternative to the existing support structure, and enough victims' families felt this way that they actually held meetings with the Department of Transportation to discuss the issue.

During these meetings in early August 1995, Secretary of Transportation Federico Peña, FAA Administrator David Hinson, and NTSB

Chairman Jim Hall met with representatives of the newly formed National Alliance of Air Disasters Association, including Doug Smith, the president of the organization, and Jenny Stansberry, the secretary. Items of discussion included the problems encountered by the families and the idea of an independent family advocate that would preclude the airlines from dealing directly with the family of the victims. The ideas were kindly received by the government representatives, but they had follow-up meetings with the airlines where the suggestions were less kindly debated.

If the families were concerned over the information flow to the airline and the questioning that occurred in the case of Eagle Flight 4184 and Eagle Flight 3379, they will be doubly concerned over recent news, confirmed by Anne McNamara, that employees of the legal department at American Airlines have volunteered for training to become CARE team members for the next time the call goes out. Dealing with a ticket agent who might be sending reports back to the company is one thing, but dealing with a trained legal expert is another, especially for a family affected by sudden grief.

With the CARE team out of the loop, the families were left to fend for themselves with the airline's representatives. Joe Bodak, vice president of field services for AMR Eagle, was the main contact with the company and was a veteran of dealing with families. Joe had been involved with the tragedies of the American Airlines DC-10 that crashed in Chicago in 1979, the 1976 accident of a Boeing 727 in St. Thomas, and now both 1994 Eagle losses. He was not a stranger to the pain felt by families. Still, decisions by the airline continued to make him a central focus of the families' anger.

This anger was the result of further perceived deceptions. Word of the field being planted instead of purchased was greeted with disbelief by every family interviewed. Each one had been assured by the airline, some reportedly by Joe Bodak, that they would not have to worry. It was a faux pas similar to the publicity disaster USAir had with its site of the crash of Flight 427 in September. That wreckage had been of a similar nature and level of destruction to Roselawn and that airline, too, had earned the wrath of the families when it reneged on its promises. That was still news when the cleanup of 4184 was going on, so nobody believed American would make the same publicity blunder.

Dr. David Dennis, the coroner, had jurisdictional control of the remains until they were released, but American Airlines had made arrangements to obtain a mass grave plot in Merrillville. One family discovered that the mass burial was to take place without notifying relatives, and word spread like wildfire. Nearly every single family was angered; part of that anger stemmed from the differing stories about who had made the decision.

It was not the first or the last time that conflicting fingers of blame were pointed. The airline was caught reeling and blamed the coroner, the cleanup company, ignorance, or "mistakes in hindsight." The more the company used the excuse of hindsight, the less anybody believed the existence of any good intentions.

The battle over the crash site heated up during the NTSB hearings in February, after additional aircraft parts and human remains were found at the site. The airline moved to quickly blame the cleanup company for not doing a good job; the cleanup company clarified the issue by stating it had only been hired to clean up the aircraft parts and not the human remains. The coroner was caught in the crossfire even though he had understood the inability to collect every scrap of remains, human or mechanical. Although the parties all blamed each other, nobody thought about what this was doing to the families.

Personal effects from American Eagle Flight 4184 were not returned to family members as they were to Lauren Anderson, the survivor of Flight 3379 in the *48 Hours* story. Several reasons explain this discrepancy. Unlike the accident near Raleigh-Durham, the destruction was extensive at Roselawn, and few intact personal effects were found; in addition, Lauren Anderson survived, and it made good publicity and a great show for the cameras.

Family members were asked to provide descriptions of valuables in the possession of the deceased and to search for sales receipts and photographs of the victim wearing these items of jewelry or clothing. Many families did not get items returned to them no matter how much proof of ownership they provided, and forcing the families to account for every item only made the circumstances more difficult for them.

The airline claimed there was no surplus of personal effects available, yet the crew working to pick up and clean the site told a different story. As with the team in the morgue, the pickup and sort crews formed an attachment to the victims. Several began talking after seeing photographs in the debris of David Shellberg wearing a special jacket and holding his horn. Although the horn was found flattened and in pieces, the jacket was intact with only one tiny tear. Workers decided not to send the items to the airline, where they knew they would be destroyed, but instead they cleaned them and sent them to David's mother. Weeks before, Rosemary Shellberg had found out David's casket contained only his hands, and the actions of the people she never met provided comfort not available from the airline.

In August 1995, American set up a dedication ceremony for the monument at the mass grave in Merrillville and made travel and motel arrangements for up to eight family members for each victim. Jim and Rosemary Shellberg each wanted to attend, but Jim could not get off work early in the week, so he needed to travel on Thursday to the ceremony on Friday, August 18. The airline made arrangements for him from Charlotte, North Carolina, to Chicago via Dallas, which meant backtracking and six hours travel when he could go direct via another airline and save time and trouble. American would not hear of the interline travel, and Jim paid $658 to travel to Merrillville on a more convenient flight.

According to Dawna Aguiar, the captain's widow, by the fall of 1995, ALPA was lobbying her not to fight the flow of the investigation or help in the campaign against the ATR. They repeatedly called her trying to convince her not to protest anything that might come out in the final NTSB report. She said she was told, "Dawna, you are out of your league on this one." The calls only served to strengthen Dawna's resolve.

Dawna also said the insurance company for ATR and American began threatening her with dredging up the couple's personal lives. They told her she had to settle with the companies and agree to remain quiet about the details, or they would bring in two people from Simmons Airlines who would testify to particular items in their lives together, such as overhearing an argument on the phone.

She also believed the airline's insurance carrier was playing games with her payments and only released her checks after prodding by her attorney. Medical benefits also were delayed or denied, causing her to at times be unable to properly provide postnatal care for her daughter, McKaila, born eight months after Orlando's death.

"I think they view me as this young, naive little girl who will let them get away with whatever they want," Dawna told a reporter during an interview designed to get the airline's attention. "They have another thing coming."

ALPA began working on the Gaglianos at the same time. In a similar fashion, Al minced no words when he told the representative that "ALPA hasn't done shit for this family or the pilots, and we are sick of it." Each month Al Gagliano read the articles in the ALPA magazine that proclaimed the helpful actions of the union and how they had been a salvation to the pilots' families. Al said it was hogwash.

The Stansberry family wanted to see the photographs used to identify Brad Stansberry. They spent a day at the office of the coroner, as did some other brave relatives, looking at more than 5000 photographs from the crash site and morgue. By the time Janie and Bruce arrived at the office that day, they believed they were seeing a pattern of deceit. They found a photograph of items collected by stake number 1015, where some remain was found that identified Brad. Included near the stake was a piece of skull bone with long brown hair attached; their son had short blonde hair. The workers tried to explain the need to mark a group of remains with one stake, to be sorted out later at the morgue, but in their distress, the Stansberrys were not satisfied. The airline finally relented and agreed to disinter the coffin with the remains sent to the Stansberrys and move them to Merrillville. A review of the records by the coroner and Steve Stits, his assistant, revealed no apparent errors, but they were unable to determine whether the skull fragment had, in fact, been put into the casket. Nobody was happy.

One report of possible mixed-up remains was never explained. Bill Readings was identified through dental records, and the coroner shipped his remains to a funeral home in Indiana. Bill Readings had requested he be cremated on his death. Without the consent of his widow or knowledge of his family, Bill's ashes were mailed to his family. The package was not certified or registered and delivery took more than four weeks. Bill missed his own funeral. After the ashes arrived, Diane Readings, his wife, discovered that the ashes in the parcel might not have been Bill's. On Christmas Eve, his parents in England also received a package of ashes that were supposed to be Bill's. Diane sent the ashes back to American Airlines.

Missy Hayes, the sister of flight attendant Sandi Modaff, who was herself a flight attendant for Simmons Airlines, attempted to return to flying after a short leave and was able to work on the Saab aircraft that replaced the ATRs during the winter of 1994 and 1995. When the ATRs came back in the spring, however, she was unable to do her job and requested another leave or groundwork. The company failed to work out the request; after causing the death of one sister and possibly the loss of another sister's unborn child, American Eagle forced Missy Hayes to give up her flying career.

Flight 4184 will probably not be the last aircraft disaster for American Airlines. The relatives and friends of those lost on Flight 4184 pray that the airline has learned from its mistakes.

11

The odyssey

By the end of February 1995, I had been through an emotional gauntlet. The third month of suspension for misconduct from my coveted pilot job after I appeared on *Good Morning, America* was ending. My reputation had been slurred by the airline spokespersons. After my disciplinary action, other pilots refused to be identified publicly, even though many spoke to reporters off the record to verify my statements. I endured a media frenzy over the crash and ATR airworthiness stories. Countless calls to my home and office requested interviews. I maintained my credibility by making the same statements time and again.

I made plans to attend the NTSB public hearings on the crash in Indianapolis at the Adam's Mark Hotel beginning Monday, February 27. The trip would also allow me to visit the mass grave in Merrillville, Indiana, and see the crash site for the first time.

At noon on Saturday, February 25, 1995, I drove away from my home in rural Plymouth, Wisconsin. My first stop was a short visit with Al and Cindy Gagliano. The Gaglianos' Kettle Moraine Ranch lies near the southern edge of the state forest of the same name. It is a large expanse of rolling hills, small lakes, and heavily forested land in southeast Wisconsin. Jeff always talked about working on the ranch and the joy it gave him to be helping his parents and enjoying nature.

I first met the Gagliano family on a frigid Saturday in mid-December 1994. I wrote to the family after I returned home from flying the day after the tragedy, and I tried to tell them how special I thought their son had been. They sent no response until early in December, when I received a card with a picture of a small white chapel and a letter inviting me to attend the setting of Jeff's monument near the chapel on the ranch. My wife and parents accompanied me to the ranch that day. As we parked the car and I stepped out, someone hollered, "Hey, the pilot's here!" A fireplug of a man in jeans and cowboy boots came from behind a building and shook my hand.

Al Gagliano grew up in a tough urban neighborhood in Milwaukee. During his youth, he was a self-described rebel and became involved with street gangs. After being blinded by a shot to the face during a fight, Al made a promise to God that if he was ever allowed to see again, he would change. He and his first wife, Jeff's mother, started this ranch, complete with an Old West-style Main Street, to give city children and individuals with disabilities a place to interact with nature, handle animals, and enjoy horse or wagon rides through the countryside. Al kept his side of the agreement.

When I arrived at the ranch this time, Al, Cindy, and I talked for an hour about new information coming from the pilots at the airline and through the families' grapevine, which was beginning to grow. Al and Cindy were leaving for the hearings on Sunday morning, and we agreed to meet for dinner in Indianapolis that evening. As I left, Cindy insisted I take a package of cookies for the road. Even though the last thing my waistline needed was cookies, I remembered another parcel of forgotten cookies and gratefully took the package. They were gone before I left the state an hour later.

I spent the night at a familiar motel near Chicago O'Hare. Walking to my car after dinner, I saw an American Airlines DC-10 pass directly overhead on short final approach to runway 14L at Chicago O'Hare. I watched until it disappeared below the buildings. My heart ached. I still missed my job as an airline pilot.

I was awake early Sunday morning and turned on the television for noise more than anything. I heard a discussion on C-SPAN regarding a front-page article on the Sunday *New York Times*. The full-screen images of the article, "Lost Chances in Making a Commuter Plane Safer" grabbed my attention.

The panel discussion and closeups of the story pages quickened my heart. Adam Bryant had come through with his promised exposé on the history of the ATR. I knew about the story; two months earlier I received a call from Keith Schneider, a Michigan writer for the *New York Times*. He had been asked to put together an investigatory piece on the ATR. I thought the article was dead until Adam Bryant picked it up in New York. I had been talking with Adam for more than a month, giving him information and explaining American Eagle operational practices as well as helping to arrange interviews with other knowledgeable parties. The *Times* had sent a photographer to my office two days earlier to get some photos of me for possible inclusion. I had almost forgotten about that session and had not expected this so soon. An hour later, I was out of the motel searching for the *Times*. When I finally found a newsstand that carried the paper, I purchased the three remaining copies.

The story was on the front page above the fold with a picture of passengers disembarking from an ATR-42 from the Simmons Airlines fleet. I scanned the first column as I walked across the street to my car. The fourth paragraph was a single sentence. "It was a crash that did not have to happen." This conclusion was reached by the writers of the article.

The story detailed the problems of the ATR on three and one-half pages, quoting experts and offering the FAA's classic line—"hindsight is always 20/20." The article discussed for the first time in a public forum the photographs of a ridge of ice behind the deicing boots that ATR showed to Howard Greene and Gary Lium in December 1987.

After placing the copies of the *Times* carefully in the back of my car, I left Chicago and headed for Indiana.

The weather was overcast on Sunday, and a cold wind blew from the north. An occasional flurry of snow blew across the highway as I passed Chicago O'Hare. The FAA had lifted the month-long ban on flying ATRs in icing conditions that it had reluctantly imposed six weeks after the crash. I knew I would not see any Simmons ATR aircraft flying into the airport because management decided to maintain the fleet in warm climates until spring.

When I reached Merrillville, I looked for Calumet Park Cemetery, the location chosen by the airline for the mass grave. Due to the extreme level of devastation, this is the final resting place for most of the victims. Nineteen coffins containing the unidentified remains of the 68 victims were interred in plots identified only by numbers scrawled on a piece of paper in my hands.

I had heard conflicting stories regarding the cemetery selection. Some families described it as a lovely, serene area, overlooking a small brook and a lake with geese. Others complained that the grave is adjacent to a drainage ditch with a crumbled concrete block wall and an old tire laying in the weeds.

After searching for 45 minutes, I finally pulled the car over to the side of the road across from a large expanse of pine boughs spread over the ground. As I walked onto the grass, the top of a small, concrete cylinder became visible. It was only 4 inches across and level with the soil, so I assumed it was a plot marker. A crust of orange clay clung in the scratched markings, and I knelt down to scrape away the earth. My head dropped as I saw the number etched in the concrete grave marker. I raised my head to see the double row of new graves, each row roughly 30 feet long and covered in mounds of pine boughs. I had found them.

After spending some time at the cemetery, I headed for Roselawn to investigate the crash site. The drive took about 30 minutes. The accident site is near the intersection of roads identified as 400 East and 700 North. The country roads are laid out in Indiana, as in most of the Midwest, on a north-south and east-west grid. I counted down the road signs, using them as a guide. Although I had seen aerial photographs of the site many times, I could not tell one field from another.

After passing through the intersection of 400 East and 800 North, I crested a small ridge. Passing a white farm house to my right, I could see the next intersection that I knew to be my destination. About an eighth of a mile south of the juncture, I stopped. To my right ran a rock road into a field and the field beyond. The pathway looked well traveled and was more significant than a farm access. At the opening in the fence, a brown wooden cross with the name Brad Stansberry on it had been placed in the earth along the fence line. Alongside the cross, several bundles of dried, windblown flowers with sympathy cards still attached were lying in the long grass and tacked to the fence post.

Not wanting to trespass, I set out in search of someone to tell me from whom I needed permission to enter the site. I stopped at the home of Norman Prohosky. It was Carol Prohosky who placed the emergency call alerting the Newton County Volunteer Fire Department that an airplane had crashed to the east of their farmhouse. Norman Prohosky answered my knock, and after I introduced myself, his eyes flashed with recognition, and he and Carol invited me into their home to talk.

I had been interviewed regarding this story at least 100 times during the past several months, but this was the first time I had to ask the questions. From the Prohoskys I learned about Clarence Hanley for the first time.

Clarence Hanley, the farmer of the soybean field where Flight 4184 crashed, doesn't own the land. He obtained a power of attorney from his elderly landlady, Mary Lutz, a nursing home resident in Pekin, Illinois. Clarence Hanley was the person I needed to get permission from to enter the land, and the Prohoskys told me how to find him.

Driving back toward the intersection of 400 East and 700 North, I headed for the white farm house on the crest of the ridge that belonged to Clarence Hanley and his wife. A knock at the door brought a burly man in his early 50s. While the exchange was personable, I sensed a tension and a wariness in his manner. I understood it—this man had gone through a lot since the accident and most likely found out that anything he said might wind up as a lead story on the nightly news. Afraid to pressure him, I touched only on the periphery of the issues and hoped to follow up later if I needed to push for any information. I still needed his permission to enter the property.

Hanley was at the site just after the accident and had led the volunteer fire department personnel to the field entrance. He had driven Bob Mauck, the first firefighter on the scene, as far into the wet field as he dared. He never walked up to the debris field, but he did see the tail section of the aircraft about 440 yards to the south of where he stood, watching Captain Mauck struggle to walk on the muddy, uneven surface as he approached the grisly scene.

The families had been told unequivocally by AMR officials that the airline was negotiating with the owner for the purchase of the land. Several families told me they had been promised this land would sit idle for years to come. The farmer who owned the corn field where United Flight 232 had cartwheeled during a crash landing near Sioux City, Iowa, after the DC-10 lost all its flight controls, had voluntarily set aside the affected area of the field and allowed the placement of a monument. On the other hand, USAir had backed out of its promise to buy the land near Pittsburgh where Flight 427 went down, infuriating the families of the victims. I had hoped the word of American Airlines was worth more. The Prohoskys had shattered that hope when they told me that Clarence Hanley was still planning to plant the ground in the spring, and I knew I had to tread softly on this issue.

"The families are interested in having the field allowed to grow wild for a number of years, and AMR has told them they were purchasing the land just for that purpose," I said.

"They asked me one time early on and I refused. We have very little farmland as it is, and we are losing more every year. Besides, they ain't making any more," Clarence stated firmly, crossing his arms and glaring at me. "Besides, why would they want that anyway? You can't get to the field without going through another and I don't want a lot of people running around back there."

"Mr. Hanley, the families all realize the extreme violence of the crash," I continued. "This was very likely the worst destruction of human beings in aviation memory, and we all know despite the recovery efforts, significant remains are still buried in the earth out there. Just like the stones that come up every spring because of weather and erosion, these victims' body parts will continue to work their way to the surface for years to come. This land is considered by the families as a burial ground, hallowed ground. The thought of continued farming and the extreme idea of a plow turning up a piece of their loved one is repugnant to these people. I think we all agree on that."

"We are not going to sell the land and that is that," he said, closing the issue firmly. "And nobody has offered me anything worthwhile to leave it idle. I can't have all them weeds growing wild out there."

Later, I learned that Hanley had demanded $2 million for the 40 acres. The land was worth about $1000 per acre. But what price do you put on a promise and easing the minds of families so badly damaged?

I asked for his permission to walk out on the land, explaining that I wanted to view the site close up, and he agreed. He advised me not to drive beyond the roadway and to take care not to get stuck. I thanked him and headed for my car.

As I reached my car door, Hanley called from the porch. He had come outside and was walking toward a 5-gallon steel bucket. "That cleanup company did a lousy job out there," he said. "I already called them back a couple of times, but I picked this stuff up around the end of December. They just bulldozed a lot of these airplane parts under the ground. I had them bring in two truckloads of dirt for every load they took from the field, but there is still lots of junk out there." He poured the parts onto the drive.

"That's what I am talking about with the human remains," I offered. "These things are going to continue to show up." I could tell by the look on his face that he wanted to hear nothing more on the subject. I easily identified several cockpit switches, an igniter plug from one of the engines, and hydraulic tubing. I picked up a bundle of wire to get a closer look.

"There's lots of wire out there and seat padding and insulation. The field has been released to me, but I still have some problems," he said. "By the way, when you go back there, you are going to see they filled the craters with new dirt and that will be pretty firm yet. The orange muck around the black dirt is where they just spread the stuff around with the bulldozer. That muck is real soft and slimy, so be careful."

The county did a good job on the roadway into the field. I had no trouble traversing the first field. The road ended, and I saw a worn pathway in the mud that was 150 feet wide at the far end as it tied into a huge, oblong area of different-colored soils. The ground had partially thawed, and I was glad I brought my boots. As I walked down the pathway to the main impact points, the earth crunched and gave slightly beneath my weight. The top layer of soil was soft and moist. I wanted to show the scene to Linde, so I panned the area with my video camera.

It took about five minutes to walk into the impact area. With the wind behind me, I did not notice the smell until I got close to the debris field. My stomach churned—the thawing field smelled of death. Ask anyone who has smelled this odor. Decomposing human remains have a smell you never forget. Even in the strong wind, the scent penetrated and lingered. This was indeed a burial ground. I cannot understand how anyone could sink a plow into this soil.

The orange muck described by Hanley was visible, and as I approached it, I found that it extended on the northern edge to a maximum width of about 30 feet. The path of this putrid, spongy earth ran along the east and north ends of the debris field. The odor was strong in this area, and I later learned that this section was heavily littered with body parts after the crash.

A flapping piece of laminated paper sticking out of a mound of dirt caught my eye. As I approached it, I saw it was a torn and tattered emergency briefing card from the American Eagle ATR-72. I carefully placed it in my oversized coat pocket. Further on, I spotted a red piece of the ATR fuselage sticking out of the soil. This, too, I picked up and placed in my pocket. With each step, I found another piece of wreckage and picked it up. I don't know why I picked up these items, only that I believed that I should. I found numbered parts, personal effects, and money. Soon, my pockets were bulging with moving parts, cockpit switches, and cockpit trim moldings. I even found an American Airlines peanut bag. There was more to recover, but my pockets were full. I investigated the rest of the field to get a perspective on the area. Hanley told me about the large piece of wreckage found beyond the south fenceline, and I walked along this area.

I was in the field for more than an hour. My hands were numb from working the video camera without gloves, and I had seen enough. I walked slowly to my car. Clarence Hanley pulled up in his truck, followed by another person in another pickup. I was concerned that he had changed his mind regarding my visit. I didn't want him to confiscate the items I recovered.

"Quite a mess, eh?" Hanley said with a smile and shake of his head.

"Sure is," I replied as I took off my coat and placed it the car. Removing my mud-caked boots, I prepared to leave. "Lots of stuff out there yet. Thanks for all your help."

I felt like he was looking at me with a suspicious eye, although maybe it was just my desire to avoid a confrontation over the debris. The man did tell me he had been treated fairly by AMR. Backing down the gravel roadway to exit the fields, I sighed with relief. I headed for Indianapolis, ready for answers to new questions.

The Adam's Mark Hotel, near the airport in Indianapolis, has a convention center, which is where the NTSB public hearings were scheduled. Waiting to check in, I found myself standing behind Ric Wilkins. Ric is president of Aviation Accident Reconstruction Services, an investigative and expert company that asked for my help working with a legal team representing nearly one-third of the victims' families. Two months earlier, I ran into Ric while checking into a hotel in Chicago for our first meeting with counsel.

"We have to stop meeting like this, Ric," I said with a smile.

"Steve, how are you doing? Just get in?" Ric said as we shook hands. "I have several others coming in, and we're going out for dinner. Would you care to join us?"

"No, thanks. I have a date with the parents of the first officer, and it has been one hell of a day," I sighed. "I stopped at the crash site and picked up a whole lot of debris. And the smell. . . ."

"The debris is to be expected in a crash of this destruction level," said Ric. "And I know the odor. There are still parts of these people buried there."

"I realize that, but the farmer told me he was planning on plowing later this spring and growing a crop this year," I said. "There ought to be a way to keep that from happening. I think it's inhuman."

"I know," he said.

It took three trips to move my luggage, computer, printer, and debris to my room. When I finished, I returned a call to Al Gagliano and made arrangements for dinner, asking him first if they wanted to see the collected pieces of N401AM.

As I waited for Al and Cindy to arrive, I carefully removed the assembled parts from my pockets and placed them on the suitcase shelf in my room. Placed touching each other, the items filled the 10-square-foot area. I had not realized I had picked up so many items. The room began to fill with the odor from the items and the earth packed around them. I placed a pocketful of dirt that reeked of jet fuel into a glass container. I was shocked that the field had been declared environmentally clean. My experience as an industrial paint contractor dealing with hazardous chemicals and wastes told me a properly remediated site would not reek of the contaminant. Out of respect and to prevent the maids from disturbing the materials, I placed a towel over the display. I washed my hands over and over and opened the window, but I couldn't get rid of the disturbing odor of death. As it turned out, I lived with that penetrating smell all week.

At 6:30 P.M., while I waited in the lobby of the Adam's Mark for Al and Cindy, I noticed a news crew with a camera set up for an interview. I knew they would be interested in what was laid out in Room 224, but the Gaglianos arrived before I could approach the crew.

The Gaglianos followed me to the room. The white bath towel covering the parts on the bench located in the narrow entryway of the room looked unintentionally foreboding. As I pulled back the towel, Cindy dropped to one knee. Tears flowed from her eyes as she alternatively looked at the gruesome display and into my face. I moved back to allow them some time to absorb what they were looking at.

"You found all of this stuff today?" Al asked, turning toward me. "I can't believe this."

"There is so much debris still out there, Al. I picked this up in about 20 minutes of walking around. The only reason I stopped picking up pieces was because my pockets couldn't hold any more."

"I just wish there was something of Jeff's found. We aren't getting anything back yet," Cindy said. "I just shudder to think what else is out there. Can we touch these? Look, Al, somebody's key chain."

"Feel free," I said. I wanted to bring them up to speed on what else I had learned today. Between identifying the pieces and answering more questions, I broke the news about the site and the farmer.

"No. That can't be true," said Cindy. "I talked with Joe Bodak at the airline and he assured me that they were negotiating for the land months ago. He promised me that the land would be purchased, set aside and no farming would take place there for years, if ever again." Her face showed her incredulity and her anger. "They lied to us again, Al."

"It is our job to force them to honor their word," I said. "These people have betrayed the passengers and crew of 4184, and now they are purposefully misleading the family members, causing more pain." This was the first time I regretted my association with American Airlines and American Eagle. I was ashamed that I had been, and at the time, still was, a part of what I now view as an unscrupulous, uncaring organization. I was angry and hurt from all I had seen and experienced. I covered the array with the towel again, and we headed for the lobby to get dinner.

The television crew was still waiting in the lobby, so I introduced myself and explained who I was. The reporter seemed to recognize me as he explained he was waiting to interview someone from the Air Line Pilots Association who had not shown up on time. I told him about the artifacts in my room, and he was immediately interested. After a call to his station, he asked to see my identification and interviewed me regarding my excursion in the soybean field and the reason for my presence at the hearings. The crew wanted some footage of the items, and we went to the room to shoot for the 10 o'clock news. CNN later picked up the report and begin airing the footage during the night.

The television crew left, and we went back downstairs. Walking up the hallway toward the lobby, I spotted Robert Martens, AMR Eagle president. He was looking down the hall at me, and I boldly raised my arm high above my head and waved. He returned the wave, and I could see him squinting to identify me. I stuck out my hand, and as he shook it, I smiled and said, "Mr. Martens, Steve Fredrick. I haven't seen you since the AMR Eagle President's Conference."

Seeming somewhat disoriented, he smiled and replied, "Nice to see you again. Glad you could make it down here."

Why he used a phrase like that for an occasion like this confused me. I released the handshake and replied with a smile, "You won't be glad after the news tonight."

Al Gagliano shook hands with Martens as Cindy looked on with a glare in her eye that could melt stone. "Let's eat, I'm starving," I said with a backward glance at the confused executive. "Boy, that felt good," I confided to Cindy and Al.

"You sure have guts, Steve," Al concluded, laughing and shaking his head.

What Al and Cindy did not know is that even I don't understand where this boldness came from. The exchange with Martens bordered on arrogance. I caught the newscast in my room after dinner and then tried to relax and get some sleep.

The NTSB fact-finding hearings began at noon on Monday, February 27, 1995. These public sessions included questioning of subpoenaed witnesses by a 15-person Safety Board technical panel, led by chief investigator Gregory Feith, a board of inquiry consisting of four senior NTSB officials, and representatives of the eight parties to the hearing. All these groups were allowed to question the witnesses. NTSB board member John Hammerschmidt served as chairman of the meetings. I had been warned by several individuals familiar with the process that these hearings would be scripted, formal, and would most likely reveal little new information.

The Hall of Champions conference room was divided into several areas. Board representatives were seated at a large table on a raised platform in the front of the room under a large NTSB emblem pinned to a dark blue curtain. The technical panel sat on a raised platform to their right, and the witness stand and a large projection screen were to their left. Parties to the hearing consisted of representatives from the FAA, ALPA, ATR, Simmons Airlines, National Weather Service, National Air Traffic Controllers Association, and the DGAC, and they occupied individual tables in the front gallery. Each of the parties had a financial, legal, or professional interest in the outcome of the investigation, and their questions reflected the positions each group found most advantageous.

The public was separated from the participants by a velvet rope. The families had a reserved seating area near the working tables. To the left of the family area was a platform containing about 30 television cameras. To the right, press tables seated about 60 reporters. More than 250 chairs were arranged in rows in the rear of the hall for the public.

By midmorning, the press arrived and assembled near the hall. I recognized some of the reporters I had been working with on the story and was found by others I had not met previously. The public relations people for the aircraft manufacturer and airline were working the reporters, and I did the same. Everyone was looking for an angle for their newscast or publication. Word was out regarding the airplane parts in my room, and I took several groups up for photos and quotes. Interviews took place in the halls and in the convention room. Every available free spot was filled with cameras and journalists. Between short question-and-answer sessions, I took note of the other players.

Jennifer Stansberry, Brad's sister, was sought after and extremely outspoken. She and her parents, Janie and Bruce, live in Anderson, Indiana, and had been active with the local media regarding their treatment by the airline. Al Gagliano was also sought out for interviews and did well expressing the pilot's perspective. At noon, people settled in for the opening statements. Al and Cindy invited me to sit with them in the family section, and I took a seat on the far left of the first row. John Hammerschmidt struck the gavel three times, and the hearings officially began at 12:10 P.M.

In his opening statement, Chairman Hammerschmidt made several key points and outlined the scope and purpose of the inquiry. No determination of liability or discussion of the rights or liabilities of private parties would take place. The purpose of the hearings would be twofold. First, the issues to be discussed, although technical in nature, would serve to assist the Safety Board in finding additional facts that would be analyzed to determine the probable cause of the accident. Second, the hearing would provide the opportunity for the aviation community and traveling public to see a small portion of the total investigative process and the dedicated efforts put forth by investigators.

The public hearings were an exercise in accountability on the part of the NTSB that it was conducting a thorough and fair investigation, on the part of the FAA that it was adequately regulating the industry, on the part of the manufacturers as to the design and performance of their products, and on the part of the workforce—the pilots, the mechanics, and the air traffic controllers—that they were performing up to the high standards expected of them.

The inquiry collected information concentrating on eight specific issues: aircraft type certification of the ATR-72 and its predecessor, the

ATR-42; the FAA bilateral agreement and certification requirements, including the known icing certification; the joint airworthiness regulations, known as JARs, for icing certification; American Eagle pilot training, especially as regards icing conditions and unusual attitudes; the corporate structure and oversight by AMR Corporation of the four individual airlines operating under the trade name of American Eagle; forecasting and dissemination of information regarding inflight icing conditions; air traffic control, flow control, and air traffic management; and FAA oversight. These issues had come under scrutiny during the early phases of the investigation, and the board wanted to gather additional information through testimony of witnesses at the hearing.

The first witness was Chief Investigator Gregory Feith, and he supported his remarks with a nine-minute graphic video reconstruction of the aircraft gyrations from several different vantage points as it plummeted to earth. Even though the plunge took only 23 seconds, the video presented segment identifications and a minute or so of the controlled flight before the first roll. The room lights were dimmed and everyone listened to Feith and watched the screen at the front of the room. The video showed a white model of the ATR-72 against a black background, its flight path identified by a series of data points portrayed by what appeared to be an elevated monorail. Several flight instruments were depicted along the right side of the screen to show what the pilots would have seen on their airspeed indicator, altimeter, artificial horizon, and directional gyro stabilized compass. Along the bottom of the video were white buttons representing flight control positions and oscillations. The cockpit voice recorder transcript was inserted in real time but was not clearly visible to the assemblage. Feith had seen this simulation a number of times before this presentation, but it was the first time this dramatic graphic display had been seen by the public.

I noticed the first couple rows of family members in my peripheral vision. Each individual was attentive as the computer-generated representation of American Eagle Flight 4184 began to move. None, I imagined, including myself, expected the impact of what we were about to see. As the aircraft began to move, we were watching from a position to the west of the flight path and stationary relative to the aircraft centerline. I knew the approximate altitude at which the out-of-control event began, and I watched for the onset. There was no way to anticipate the effect of witnessing the recreated ordeal.

People gasped as the first roll excursion began. Rapid in its onset, the initial roll was followed by a complete rotation, and the aircraft continued onto its back as the nose fell and the ATR sped earthward.

Someone murmured, "Oh, my God!" Thankfully the graphic did not continue to the point of impact. People were crying, their thoughts filled with the perceived terror of their loved ones. The room was quiet except for muffled sobs. My head fell forward, and I looked at the carpet, my eyes filled with tears. No matter how many times I read the press accounts, I was not expecting this. As a pilot, I now understood what these flyers had to deal with and how short a time they had available to them to sort things out. Several more perspectives were displayed, but the emotional wallop of the first video image was unequaled by subsequent images. Most of the families now knew and understood more than they wanted to. Nightmares haunted some for weeks.

All week, during the testimony and breaks, my time was spent answering questions from the families and working the press. I thought it was important to do both, but during the first break on Monday afternoon, I experienced my first unexpected and wonderful encounter. A lovely, middle-aged woman approached and spoke to me in a low voice that expressed a desire to keep the conversation away from prying ears.

She was looking me in the eye as she asked, "You are the pilot, aren't you?"

"Yes, I am," I replied not knowing what to expect.

Her arms stretched out to me and as she moved to hug me, she confided, "I lost my son on this flight. I wanted to thank you for what you have done."

"I am so very sorry for your loss," was the best I could do, returning the hug. I was at a loss for words but deeply touched.

We held a short conversation, but the sound of the gavel called us back to our seats. I thought of this lady frequently through the week. She would never know how much her expressed support of my actions meant to me until now.

On Monday I also met Tom and Terri Severin. Terri's sister, Patricia Henry, and Patty's four-year-old son, Patrick, died in the accident. The Severins were at this time caring for Patricia's older son, who was on a United flight to Chicago O'Hare at the time Flight 4184 crashed. Tom and Terri visited my room to view the display of aircraft parts.

Three meterological experts were questioned on Monday. The first witness, Marcia Politovich of the National Center for Atmospheric Research in Boulder, Colorado, seemed most interested in making a pitch for additional government funding and complaining about recent budget cuts. The second witness, Professor John Marwitz of the Department of Atmospheric Science at the University of Wyoming, dazzled the audience with his new discovery of "drizzle drops," which were not exactly rain but not exactly drizzle either. The professor also compared the icing accumulated and tolerated by a King Air business aircraft in the Rocky Mountain range to that experienced by airliners in the midwestern United States. The Great Lakes area, with its copious quantities of fresh water, has generated some of the worst structural icing on aircraft in the world. Professor Marwitz made a number of statements about the aerodynamic conditions that seemed to defend the ATR position. The final witness of the day, Melvin Matthews, supervisor of the National Aviation Weather Advisory Unit, was most interested in the development of algorithms to numerically define the atmosphere.

The gallery experts categorized the day's progress as less than minimal. The hearings adjourned for the day at 6:59 P.M. My day ended at midnight when I finished reviewing the documents obtained during the day.

On Tuesday morning, Missy Hayes, the sister of flight attendant Sandi Modaff and a flight attendant for American Eagle herself, came to my room to see debris I had gathered, accompanied by her husband. Tuesday afternoon, the Severins and the Hayeses made a trip to the crash site.

On Tuesday morning, the testimony of FAA air traffic control officials revealed that Flight 4184 was held because of the high volume of jet traffic inbound to Chicago O'Hare from the west. It was inconvenient to work the turboprop commuter aircraft into the flow. Weather and strong winds at the airport had precluded the use of runway 9R, thereby reducing the number of arrivals.

Unfortunately for the crew and passengers of Flight 4184, the flight was handled by a relatively inexperienced controller. Veteran controllers at Chicago O'Hare are familiar with the capabilities of the ATR-72. At altitudes of less than 10,000 feet, where all aircraft are limited to a 250-knot maximum speed, this turboprop airliner can perform as well as any jet. In fact, the ATR might be more capable than the jetliners at that speed. Many times I have flown at the maximum speed of 250 knots to the outer marker, about 5 miles from the airport, slowed and configured the aircraft, landed and minimized my time on the runway by exiting at the first available high-speed taxiway, something a Boeing or Airbus aircraft cannot do.

FAA policy requires the filing of a report if a flight is delayed for longer than 15 minutes; however, a loophole in the system allows controllers to add additional 15-minute chunks of time to delay the hold without the penalty of additional paperwork. Flight 4184 was issued a 12-minute hold, then a 15-minute extension, and finally a second 15-minute extension for a total hold time of 42 minutes. Ironically, had the controller issued a 42-minute hold to begin with, the crew probably would have remained at a higher altitude to burn less fuel, maintain a smoother ride, and, as evidence suggested, stay out of the dangerous ice present at the lower altitudes. The decisions made by the crew definitely would have been different had the controller been more concerned with the safety of flight than the additional paperwork.

The board probed for weak spots within the FAA. However, when a soft or deficient area was touched on, the interviewer did not seem fully prepared and failed to ferret out important information. It seemed as if government witnesses had been coached to only offer what was asked and to defer if not 100-percent certain. The hearings took on a positioning tone rather than one of fact-finding.

The first witness in the operations area was Robert Zoller, vice-president of operations for AMR Eagle. Zoller shed little light on the causal portion of the crash and spent most of his time defending the organization and interaction with the FAA and the four independent airlines.

I had not noticed the witness list posted on the main bulletin board, so I never expected the inquiry to move in the direction it next went. As Zoller finished his testimony, Captain Dan Rodts walked into the room, escorted by Edward Harvey, the Simmons Airlines director of flight operations. Captain Rodts was the captain of the aircraft I had been flying on the night of our icing-induced control anomaly during an attempted landing at Marquette in November 1993. The investigation was indeed taking an unexpected turn.

After I had gone public on the safety issues, I had asked Dan for support with the reporters regarding my statements of fact. His response was supportive, but he indicated an uneasiness at being associated with the media. His airline job was the total support of his family, and he could not risk his career. I understood Dan's reluctance and never held any ill will toward him. Others were speaking anonymously to the press to open the doors to the problems with the ATR. When he walked into the room that day, I was surprised. This particular captain was always calm and collected; however, this day he appeared nervous and uncomfortable. Worst of all, he would not make eye contact with me. Even when I moved to a sideline position and was a mere 15 feet away, Dan did not acknowledge my presence.

I remembered the conversations we had after the accident and how dismayed he was about the lack of response to the aircraft's problems. Dan even confided in me how he had lobbied the flight operations supervisors who were responsible for safety regarding the need to dispense icing operational information two months before the accident. He expressed his frustrations regarding the mishandling of our incident more than a year ago. Dan knew nothing would change without dramatic events, but he had to remain on the side-lines, safe from the wrath of the company. I could not find it in my heart to resent his inaction. I saw his inner struggle.

When Dan took the witness stand, my heart raced. Would he disavow his previous concerns and throw in with the airline? Or would he speak the truth? At this point, I don't think even he knew.

After some preliminary questioning, the interview moved to the Marquette incident. I sat forward in my chair, straining to hear every word and pick up every intonation. My credibility was at stake, but more important, the entire issue of inaction associated with the checkered history of the ATR could be dealt a severe blow if Dan crumbled. He started to discuss the November 1993 event in a nervous but even tone.

We were operating an ATR-72. It left Traverse City, Michigan, and initially we left Chicago for Traverse City. It was an intermediate stop. At that point, our continuation was to Marquette, Michigan.

My first officer and myself reviewed the weather that came across with our release. We reviewed the forecasted weather, the current weather, and we discussed the situation that we were about to go into. And based on the weather that was in the area, it was overcast skies, visibility was roughly two miles, light snow showers, and there was also a remark in our weather packet that there would be a good chance for a possible temperature inversion.

For my first officer and I, that would indicate that there was also a good chance for icing. We had an indication of where the tops of the clouds were in the Great Lakes area from our descent from Chicago into Traverse City. And we discussed the fact that going into Marquette we would be getting some ice.

We left Traverse City. En route to Marquette, we were on top in the clear. And going into Marquette—it's an uncontrolled airport. It sits right close to Lake Superior. The terrain is hill, not—they call it somewhat mountainous, but they are big hills. So it's just an unusual airport to go into.

So we were mentally prepared for it. We had discussed what we were going to get into. We picked up the weather through the automatic weather information that we can receive through our radio, and we got the weather report from that particular source.

We also received weather from K.I. Sawyer, the approach facility. I also received company weather, just because we wanted to get as much information as possible.

We went into our descent for Marquette for an ILS approach to runway 8. That was the prevailing runway. The winds were suitable for that runway. And the conditions—due to the overcast skies and the visibility, that's the runway that we elected to land on.

Prior to entering the tops of the clouds, we selected level 2 anti-ice. Now, we selected this in anticipation of the temperatures being just right for the use of the icing level 2 and also to make sure that everything was operating properly. So that was selected before entering the clouds, the tops of the clouds.

We also selected level 3 prior to entering the clouds, because with my judgment, my experience, I've been in this airplane for a long time, I knew I was going to get ice. I want to make sure this equipment is working. So level 1, 2, and 3 icing equipment was working at this time.

We got into the clouds. There was a slight bit of turbulence going into the clouds. That's normal. We were getting vectored for the approach. I believe we descended to 3600 feet. As soon as we got into the tops, the ice detector went off. The first officer and I had confirmed we were getting ice at this time. The icing equipment was working. It was functioning. Everything was fine.

We were thoroughly prepared for the approach, due to the conditions that it was an uncontrolled airport, we knew what we were getting into, and all we wanted to do was fly our airplane, monitor the approach, take care of the necessary calls to our company, to unicom. Just in case there was somebody else out there on the field, snowplows and other aircraft, whatever.

So all this was done. And at this time, all we had to do was concentrate on flying our airplane, monitoring the icing that we were getting, and making sure all the navigation aids were functioning properly. This was all done.

Prior to the final approach fix, the aircraft was completely configured for the approach and the landing. Everything was done. We made a call prior to the approach fix to K.I. Sawyer with the remark and a pilot report for moderate rime ice in the clouds.

Prior to this, we also increased the prop RPM to 92 percent. This was done because we did not want to be distracted going into this airport. We wanted to concentrate on flying the approach, and we did not want to get any distractions. The props were previously set at 86 percent.

The reason why we set them at 92 is because there are oc-casions where you can get a prop vibration that would be a distraction. To alleviate this, we ran the props at 92 percent.

This airplane was configured. We were all set and ready to just fly the ILS approach. We crossed the outer marker. First Of-ficer Fredrick was flying the airplane. The autopilot was off, and he was hand-flying. The approach was stable. It was smooth. We occasionally received a light chop going into Marquette that night. It was very calm, smooth air. As much as I had flown up there, it was a very nice night going into Marquette.

I constantly monitored the icing situation for the first of-ficer and kept him abreast of how the accumulation was—at the rate it was accumulating. And also I made note that the deicing boots were working properly. We were breaking the ice. We were shedding the ice. It was normal.

Inside the final approach fix and on the descent, on the glideslope, on the localizer, this approach was wired. I mean, he had it perfect. There was no deviation from the glideslope. The airspeed was fine. It was textbook perfect.

Then, on the descent, inside the marker, I noticed on the ice evidence probe, which was on the first officer's side at that time, the icing was rime. It was conforming to the icing evi-dence probe. The shape was the same as the probe.

And in a short period, it started to change. Where it wasn't conforming, it was starting to protrude on the upper part of the cylinder and the lower part of the cylinder. I made a comment to my first officer. He looked at it. He was able to see it. We made a comment with regards to that, and we mon-itored the approach.

The boots and the deicing equipment [were] functioning. The ice was being shed at that time and we're inside the marker. All of a sudden, I felt a yaw, a slight yaw to the left, and I thought possibly we lost some power on the number one engine. I had made a comment to my first officer, because the flight was extremely coordinated.

There was no out-of-coordination as far as rudder, as far as any of this. And I confirmed that, no, we did not lose power. And I made a reference to the first officer, I asked him, maybe we should put in a little rudder trim there, and he did.

All of a sudden, I felt a slight roll to the right. The aircraft began to settle a little bit. The speeds were normal. And all of a sudden, at that point, when all this happened—and mind you that this happened extremely quick. It's not as slow as I'm talking right now. It yawed left. It rolled slightly to the right. And all of a sudden, we got a lot of vibration coming from the tail section of the airplane. It was very minute. And by the time it got to the cockpit, it was like a crescendo. It was extremely loud.

By that point, I had my hands on the throttles. We went around. We executed a go-around. The first officer pitched up between 5 and 10 degrees. The airplane was flying. We were going straight ahead, and we received an audio alert.

At that time, I did not know if it was because the control column was shaking pretty good. It was shaking severely. The aircraft was vibrating severely. At that point, I could care less what sounds I heard. I was flying. I was climbing. I was going straight ahead. I was managing the situation. I was making calls to my first officer.

We both, at that point, thought it probably was an overspeed warning. We made reference to our airspeed. One call was for 135 knots. The other call was for 137 knots. We continued to climb at 140 knots. And we disregarded the audio alert, because at that point, we knew where the tops were at. The were at roughly 5000 feet. All we cared about was that we could fly this airplane, climb our heading, get on top, get our composure, and take care of business. And that's exactly what we did.

Once we got on top, we regained our composure, made the appropriate calls to K.I. Sawyer. We called our dispatch. We elected to go to Green Bay, Wisconsin. Through dispatch, we got the current Green Bay weather. It was in our weather packet also. We elected not to go back into Marquette. We elected to go to Green Bay.

The weather was 33 degrees in Green Bay. It was IFR, but there were no reports of icing. There had been other aircraft in the area, and also that was the best course of action.

We landed at Green Bay, Wisconsin. We had passengers on board that needed to be attended to. The first officer and I got out of the aircraft at the passenger exit. We looked the aircraft over visually from the ground. We saw unusual ice formations on the aircraft.

The first officer, after a period of time, was able to get up into a deicing truck, which has a cherry picker, which is a bucket that can be operated to get up into the tail section or elevated places around the airplane. He went up and did an inspection, wrote a report, described the icing, and I notified our dispatch when I was on the ground in Green Bay.

I notified our maintenance personnel and had talked to them about the entire situation, both dispatch and maintenance. That's basically about the extent of that.

I began breathing again. Dan had accurately relayed the event, without emotion and without sensationalizing or downplaying the occurrences. Although he did not voluntarily advance any comment in further questioning regarding the handling of the subsequent investigation and did not make any reference to the other peripheral issues of training and management inaction, he had told the truth. After all the time that had passed, the only other individual who knew exactly what had occurred that night over Marquette had finally ended his silence.

After the hearings concluded that day, I was surrounded by reporters wanting to know what I thought of Captain Rodts' testimony and if I felt vindicated. I smiled and told them the truth. "He did it right."

As I headed back to my room, I was flagged down by a representative of the flight attendant's union, who asked whether I would speak to Amanda Holberg's mother. It was now after 8:30 P.M., but I wanted to meet this woman and answer any questions. During our conversation, I learned a little about Amanda Holberg. She sounded like the type of person most of us would remember even after a short encounter. Mrs. Holberg maintained an inner strength that amazed me.

Back in my room, I noticed my message light flashing on the telephone. Tom Severin had called and wanted me to get back with him tonight. I remembered they had been out to the crash site, and I was curious to hear what had happened.

The phone in the Severins' room was busy, so I left a message. Several minutes later, Tom Severin called. He told me about the excursion to Roselawn, about meeting with Clarence Hanley, and about wandering around the field. Tom informed me they found what they believed to be body parts lying in the mud, uncovered by a recent rain. Incredulous, I asked him what they did with the remains and was stunned to learn that they were in the Severins' motel room to be picked up in the morning by Dr. David Dennis, the Newton County Coroner.

"Tom, how did you find these things and what do you think you have?" I asked.

"We were walking around the field, finding parts of the airplane lying all over, and suddenly I just spot what appears to be a bone sticking up from the mud. So I call Terri over, and we pick it up and put it in a bag we had brought along."

"And what exactly did you find?" My mind was racing. This was exactly why we had lobbied for setting the land aside. Remains were still there.

"I came across three items. One was the bone fragment which appears by the size to be from a femur, the large thigh bone. Then I found a mass of bone and sinewy meat and nerves that could be from the lower lumbar area, and then there is a piece which may be a heel bone or it could even be a piece of wood, but the other two I am sure about."

"What about the press?" I asked. "Have you given any thought of strengthening our position on the land by talking to the media?"

"Do you think it would help?" he asked. "Would there be any interest? I don't know if Terri and I want to have a media circus downstairs at the hearings. Dr. Dennis is coming by to pick the items up in the morning, and I don't think it would be fair to ambush him like that."

"I'll tell you what," I said. "We could have a single news outlet get a scoop and some pictures of the items in your room in the morning. That way, you just quietly turn the parts over to Dr. Dennis and we still can get the advantages of the exposure," I suggested.

"Let me talk it over with Terri and I'll call you back."

"Can I at least start the ball rolling and see if there is any interest without using any details? It is already after 9 P.M., and we need to get people going. And I would like to tell Al and Cindy Gagliano."

"That's fine," Tom said, and we hung up.

After hanging up, I called Michelle Kelley at CNN. I had worked with her before, and I knew that would be the best possible use of a single media outlet. Michelle was interested in the exclusive, and I promised to get back to her shortly. I also contacted Al and informed him of the discovery. He was appalled, and Cindy was irate.

Several minutes later, my phone rang. Tom Severin gave the go-ahead for the media. The only conditions were that we do it before 7 A.M. and that we do not do it in their room. I told him I would work out the details and asked them to come to my room between 6:30 and 7:00 A.M. on Wednesday.

On Wednesday morning, I was ready for the knock at my door precisely at 6:30. Tom and Terri entered my room with their macabre discoveries. Inside a small red and white cooler were the partial remains of two crash victims. Al Gagliano arrived at the room shortly after the CNN crew, led by Michelle Kelley. Cindy Gagliano elected not to come.

The day was overcast, and the sun had risen. The room was lit only by the few lamps in the room. The camera light came on, and Tom removed the bags containing the pieces and laid them on the table. They had washed them slightly the previous night to remove the dirt. With every bit of dignity they could muster, the group examined, turned, and photographed the remains.

Tom and Terri Severin, until now having refused to talk to the media, were thrust onto center stage and did a tremendous interview with CNN. The reluctant couple spent most of the day discussing the findings with reporters and press.

The hearings were scheduled to begin again at 8:30 A.M. The television cameras were again in place on the platform. I found Jenny Stansberry and several other family members and warned them about the findings and the media hype that would most certainly take place. Then I stepped back to let the families do as they wished.

John Hammerschmidt pounded his gavel to open the hearing. Chairman Hammerschmidt was one of the last to learn the latest dark secret. By afternoon, the airline and PWI, the cleanup company, were issuing press releases, each blaming the other. The governor of Indiana ordered another search of the field, and in the town of Roselawn, the heroic rescue workers unjustly took the blame. It was not a matter of who is at fault, because how can you fault an impossible task? It was a matter of human dignity and kept promises. It was a matter of feelings versus profits. We felt we might have stopped the farmer's plow.

At the hearings, the weak questioning continued when representatives from ATR were called to the stand, leading Al Gagliano to approach a representative from ALPA and ask why they were not pushing for answers on the aircraft safety and performance problems exhibited in some types of ice. The response was short and to the point.

"Mr. Gagliano, what you don't understand is how this is all related," the official declared. "Before the hearings, a backroom agreement was reached between ALPA and ATR whereby the pilots' union would go softly on the aircraft performance and the airplane manufacturer would tread softly on the pilots' performance."

"But what about getting at the truth?" Al asked. "Are you guys here to do that or just to cover for one another?"

The response was a shrug of the official's shoulders.

On Wednesday afternoon, the media attention on the remains found by the Severins brought David Allison, president of PWI Environmental, to the hearings along with a press release on the site cleanup. His firm was hired by Simmons Airlines to perform certain functions at the site. Although his personnel were to clean up aircraft wreckage and remediate the fuel and oil spills in the soil, they were not involved in the recovery of remains other than as support and suppliers of protective gear.

The airline, on the other hand, issued a statement that said, "At the state's direction, AMR Eagle contracted with a state-approved biohazard clean-up firm, PWI Environmental, to conduct the site clean-up to the specifications and satisfaction of Newton County and Indiana state government authorities. The crash site was under the control of the National Transportation Safety Board (NTSB) and the Newton County Coroner, the Indiana State Emergency Management Agency and local emergency officials, not AMR Eagle."

Even Peter Piper, then president of Simmons Airlines, was quoted as saying, "After being billed several hundred thousand dollars for this clean-up, we are very disappointed with suggestions, assuming they can be corroborated, that the site was not thoroughly cleaned."

Who was to be believed? We might never fully understand what transpired in the recovery effort. PWI has gone to court to collect on its contract with Simmons Airlines. Note that Mr. Piper's statement did not mention making a payment for services.

In my discussion with David Allison of PWI, I learned that the airline used its clout with government authorities, who had in turn ordered PWI to release the aircraft wreckage back to AMR. American had begun destroying the wreckage, which was possible evidence if the case ever went to trial, and was denying experts a chance to view and investigate the remains of the ATR-72. Fortunately, according to Allison, because AMR had also not paid the disposal company for its work, the company had stopped the destruction.

After learning this information, I found one of the attorneys for the families and informed him of the destruction of possible evidence. I knew that once a lawsuit was filed, the parties have a responsibility to preserve evidence, and in this accident, aircraft wreckage would definitely be evidence. Lawsuits have been filed in a number of jurisdictions and served on AMR. The associate quickly contacted his superiors, and an emergency hearing was set. The court issued an order to preserve the remaining wreckage.

After the hearings were over Wednesday, I walked back into the hearing room to talk to Al and Cindy Gagliano, and I noticed a petite, pregnant woman standing behind them. It was Orlando Aguiar's widow, Dawna. I knew they had been expecting their second child at the time of the accident.

Al introduced me to Dawna Aguiar, and the first thing I noticed is how fragile she seemed. She cannot be more than 5 feet 2 inches tall and is slightly built. Dawna was only five months pregnant. A protective pilot friend accompanied her at the first couple of days of hearings, but I was so touched by the woman standing in front of me that I did not remember his name immediately after I met him.

In the few moments of conversation we had, I sensed Dawna's overriding concern was not for herself but for the other family members. She was here to bear witness for her husband. Despite her frail outward appearance, I learned that this woman, with one daughter almost two years old and her second child due to arrive in June 1995, was one of the strongest personalities caught in this quagmire. I grew to respect her tremendously.

By the end of the testimony on Thursday, which included engineering and test pilot experts from ATR, the positions of each of the potentially liable parties had been identified through the points they deemed important to place on the record. ATR was hiding behind the FAA's aircraft certification requirements for icing approval and attempting to ensure testimony specific to Appendix C of the Federal Aviation Regulations was entered.

Appendix C, written more than 40 years ago, involved the definition of a narrow band of weather phenomena in which an aircraft must prove performance to achieve certification, and thereby approval of the FAA, to fly in known icing conditions. What Appendix C neglects is the comparison of the old-type, thick airfoils in use four decades ago with their forgiving capacity to carry ice and the new higher-technology airfoils that perform poorly when the airflow is disrupted. The NTSB has been lobbying the FAA for many years to change Appendix C to address the advances in technology and weather knowledge.

In media interviews, the public relations team for ATR attempted to cloud the issue of performance in Appendix C by issuing statements such as "we have recently tested this aircraft and will test the new, larger deicing boots in conditions four or five times the FAA requirements." This statement referred to testing in water droplets greater than 200 microns, compared with the 50-micron limitation in Appendix C. Although mathematically correct in that 200 divided by 50 is 4, it is a disingenuous and misleading representation of the science, according to the most respected icing authority in the United States and one of the authors of the original Appendix C.

The size of a water droplet is only one factor in a complex and changeable series of factors that affect the potential of the droplet for icing. Objective scientists understand that the earth's atmosphere is not a homogeneous substance. On the contrary, the atmosphere is in a constant state of change, with boundaries between similar but not consistent air masses moving in three dimensions. To attempt to simplify this ever-changing sea of air surrounding our planet in only one term seems ludicrous.

The airline appeared to be using the excuse of only doing what everyone else told it to do. The manufacturer and the FAA are there to issue the directives with which the airline is to comply. The lack of substantive questioning of any witness by ATR and the FAA made me wonder whether they were participants in or observers at the hearings.

The FAA needed no excuse. Its employees in charge of various operational or technical areas are either not responsible or do not know. How can the person charged with overseeing the safe operation of the airline be so powerful yet so impotent? Jerry Barron was the principal operations inspector responsible for Simmons Airlines. He had the authority of the FAA standing behind him and the trust of the flying public looking to him. Yet he chose to ignore compelling evidence pointing toward a potentially critical safety issue.

I met Jerry for the first time at the first officially unofficial meeting after the incident in Marquette. I was shocked to see him again standing behind me in line at the restaurant during lunch shortly after his testimony. He told me, "ATR classified the Marquette incident as a non-event."

"Jerry, how can you allow them to classify our control anomaly as a 'non-event'?" I asked.

"I never said that. That was ATR's comment," he stammered.

"I know what you said, Jerry. I was there. The question is how can you allow that statement to close the inquiry when you know the facts?"

"I don't know," he responded and shrugged.

When the personnel from NASA Lewis Research Center testified on Thursday afternoon, the aerodynamics of what occurred were shown by a slide of a cross section of the ATR-72 airfoil in the wind tunnel. A dramatic flow separation occurred when the ice built to a ridge on the unprotected wing surface. This separation occurred at low angles of attack and caused a "sucking" or snatch of the aileron as seen in the data recorder readout from the accident aircraft and other incidents.

Interestingly, the larger the drop size, the smaller the buildup or elimination of any ridge of ice. The largest ridge formed with moderate-size water drops, those closest to the Appendix C standards; however, droplet size was only one of many factors. The development of the ice shapes at NASA, being given the weight of a proven theory, completely negated the relevance of testing to four or five times the certification standards held out by ATR.

FAA experts testified on Friday that the results of the NASA research would allow them to force ATR to comply with the expanded standards in the interest of "continued airworthiness." The concept of "continued airworthiness" means the aircraft must maintain its airworthiness for the life of the airplane type. This, in essence, gives the FAA authority to issue an airworthiness directive to deal with this recent safety issue and force the operators of the aircraft to comply with special rules outside of normal certification criteria.

The hearings concluded and recessed at 6:27 P.M. on Friday, March 3, 1995. It was the hardest week I have ever experienced. I spent the night in Indianapolis and drove home Saturday. During the drive, I resolved to write the book that first suggested itself while I soaked in that hot bath in Montreal.

12

In harm's way

One of my favorite films is *In Harm's Way*, a John Wayne classic. The most poignant scene takes place several days after the surprise attack on Pearl Harbor. As John Wayne walks Kirk Douglas to the harbor boat, they see a new warship going out of the harbor.

"The new cruiser going out to join Halsey," says Douglas. "She's a tiger."

"A fast ship going in harm's way," replies Wayne. "A lousy situation, Commander Edington."

The philosophy John Wayne expresses at that moment in the movie corresponds to how I feel about the ATR aircraft in some icing conditions. The aircraft is notorious in the United States because people have become aware of the aircraft and the circumstances surrounding the tragedy of October 31, 1994, but they are aware of only part of the story. Most people believe the icing problem is fixed because the FAA held a press conference and told them it was. But ATR aircraft continue in harm's way even as I write this chapter during the early fall of 1995.

Because ATRs are certified in the United States under a bilateral agreement with the French government, the FAA relies on the data authenticated by its French counterpart, the DGAC, and participates only as an oversight authority. The FAA has the final word on the issuance of a type certificate for airliners built outside the United States, just as the DGAC holds the same power over aircraft built in the United States, but political and financial realms often work together to exert pressure on government agencies.

The standards for certifying a transport category aircraft are similar, but not identical, in both countries. The French maintain that their icing certification standards are based on and meet the Federal Aviation Regulation (FAR) Part 25.1419 and Appendix C, the icing standard used for all transport category aircraft manufactured in the United States. Early certification documents, however, showed that

ATR repeatedly sought exemptions or special treatment on various issues from the FAA. An extremely contentious area of disagreement was over the aircraft's ice-protection design and efficacy. In the end, even though the ATR got past the FAA oversight during the initial certification process, the system was designed with ongoing backup procedures for problems.

A postcertification method of regulation used by the Aircraft Evaluations Group (AEG) of the FAA, located in Seattle, Washington, is "continued airworthiness," under which an airplane can, at any time, come up for additional review and certification modification or revocation if any issue arises to call its airworthiness into question.

On January 11, 1995, the FAA lifted the ban on flying the ATR in known or reported icing conditions. At a news conference held in Washington, D.C., speakers, including FAA administrator David Hinson, proclaimed a technological and informational breakthrough that now allowed the aircraft to fly in ice.

The news conference was made possible by intensive closed door meetings held among the FAA, ATR, the airlines, and the pilots' unions, including one on December 30, 1994. In that meeting, Anthony Broderick, associate administrator for regulation and certification, led participants into an informal agreement that allowed the NTSB to deliver a press release on that date.

Pat Cariseo of the NTSB phoned me to relay the news of the meeting and the ATR proposal to get the aircraft back flying in known or forecast icing. As the news release stated, the proposal was "to retrofit the wings of ATR-42 and ATR-72 aircraft with improved de-icing boots that approximately double the amount of the wing area protected from ice." The plan allowed for approval of the device by U.S. and French authorities, with the spring of 1995 as a target date for the completion of the work on the 175 U.S. ATRs. The second part of the proposal called for the introduction of new dispatch and operating procedures while the ATRs were being modified, as well as additional pilot and dispatcher training.

Pat Cariseo did not tell me everything that occurred at the meeting on December 30th. When I later reviewed the meeting notes, they contained a handwritten message relating to the current poor public perception of ATR's product. An FAA official scribbled the notation, "ATR wants FAA to help build public confidence." A foreign manufacturer was requesting that overt sanctions be provided by a U.S. federal agency.

ALPA, the pilot union, had a problem with its position on the ATR ever since it drafted letters to the FAA protesting the "Band-Aid fix" proposed after the 1989 icing instances. Seemingly dormant since then on the safety of the airliner, ALPA announced its support of the aircraft after the accident and became especially vocal in that support after public attention was focused on problems beginning in the first week of December 1994. After the FAA announced the grounding of the aircraft on December 9, 1994, ALPA withdrew its support of the ATR when only one day earlier it had been publicly and vigorously defending the aircraft and downplaying media statements by myself and others. When it became obvious the airline would move crews to stay with the aircraft during the flight restrictions, ALPA began defending the grounding to its membership and proclaiming its active role in the suspension of icing operations in the interest of safety. The Simmons Airlines ALPA hotline provided supporting evidence with a statement that revealed the control forces were estimated to be nearly 300 pounds during the rapid descent and noting that even with both pilots fighting to save the aircraft, they would not be able to overcome that type of control pressure. The ALPA hotline continued to publicize the union's strong sentiment against the lifting of the ban, citing continued safety concerns, until just three days before the January 11th announcement. Simmons ALPA master executive counsel (MEC) chairman Robert Krzewinski stated that the unions of ALPA; Allied Pilots Association (APA), which represents Flagship and Executive airlines of the American Eagle family; Regional Airline Pilots Association (RAPA), which represents Wings West, the last of the Eagles; and the Canadian Airline Pilots Association all had listened to ATR's pitch and were not buying it. Somehow, on January 11, ALPA national president Randy Babbitt sat on the dais with the FAA and proclaimed the ATR safe once again, "if operated under the newest of the rules."

ATR protested the estimates by the engineers on the control forces required to regain control of the aircraft after a roll anomaly caused by airflow separation. They insisted the ATR still met the FAR Part 25 guidelines of a maximum of 60 pounds of control wheel force with nonhydraulically powered flight controls. The numbers might never be entirely resolved, but when ATR got a control wheel force of 60 pounds, the SAAB 340 had a control wheel force of 15 pounds during the same anomaly scenario. The 15-pound force for the Saab was determined during ATR's own testing on one of the competition's aircraft leased from an airline by ATR.

Charlie Periera of the NTSB recorded meeting notes indicating ATR had researched the feasibility of placing hydraulic controls on the ATR after studying an incident in 1988. ATR had considered hydraulic controls to aid the pilots in similar situations, but they apparently discarded the idea because it was too costly and would cut into the company's competitive position. Even though ATR later protested the meeting notes and demanded the removal of the contested item, Charlie Periera held firm and provided a complete copy of his notes to the *New York Times*, against the wishes of the NTSB, which wanted only an edited version released.

An airline announced in the meeting that it would be able to stay afloat only two more weeks and would shut its doors if the ATRs were not back flying full schedules by mid-January. American Eagle had a wide system that allowed shuffling of the aircraft between cold and warmer climates and had already begun redispatching ATRs and their pilots. Other airlines, such as Trans World Express in St. Louis, did not have that luxury, and for most of a month, its ATRs sat idle. Some attendees at the meeting even stated that ATR had threatened the shutdown of its manufacturing plant.

In addition, at the meeting ATR announced the discovery of a freezing precipitation not previously known, which had droplet sizes between that of freezing drizzle and freezing rain. The new discovery was called super-cooled drizzle drops, shortened to SCDD. Unfortunately for ATR, the people defining these droplets also indicated in a press release that they first knew of SCDD in 1982, 12 years before the accident.

Another portion of the new information involved the trapezoidal ice formation on the side window of the aircraft when operating in conditions outside the certification envelope. ATR determined this to be a new visual cue to determine the presence of dangerous freezing precipitation that would notify the crew to exit the area before trouble developed with icing behind the deicing boots. ATR pilots at Simmons had used that side window ice as an indicator for years, and even though we had bad ice during the Marquette incident in 1993, we did not observe side window icing. The same type of ice developed on Flight 4184 on October 31, 1994, but apparently its pilots also did not observe side window icing. In more than five years with 3700 hours of flying the ATR in the upper Midwest, I have only seen the side window ice on two occasions. ATR, however, proposed this as a visual cue after its test pilots reported it during 18 hours on six test flights behind the Air Force KC-135 icing tanker flying over the California desert.

The third and final new discovery was the promise of the new deicing boot. The enlarged boot was based on research done after the Roselawn tragedy, although a larger boot was recommended by experts after almost every incident, including the Como, Italy, accident in 1987. Each time, ATR disregarded the recommendations, even though its own testing had proven the possibility of ice behind the original boot. The manufacturer promised the new boot would cover 16 percent of the wing's chord, although because of the limiting factor of the current removable leading edge of the wing, the boots would be shrunk to 12.5 percent of effective boot area. The difference was critical, because the Bascombe-Downs testing in 1988 had proven the impingement of Appendix C droplets to 23 percent of chord, and even ALPA's technical document submitted in qualified support to the NTSB during May 1995 indicated the Roselawn accident ice ridge had, in all likelihood, developed between 8 and 16 percent of chord. This meant, at the reduced boot size, 3.5 percent of possibly the most dangerous area for an ice ridge to develop was still left unprotected. However, because of the original design of the aircraft, it was the most cost-effective route. ATR issued a statement on December 10, 1994, after the grounding: "ATR understands that DGAC feels no further testing is needed, and the measures adopted are sufficient."

Anthony Broderick ran the press conference on January 11, 1995, after opening statements by his boss, David Hinson, and ALPA chief Randy Babbitt; he laid out the new operating rules that basically rehashed the old operating rules. Unfortunately, the press did not pick up on the sleight of hand or the favor the FAA did for ATR in the text of the official airworthiness directive.

The FAA refused to interfere with the NTSB, stating that the official cause of the accident had not been determined. However, the FAA gave the ATR a vote of confidence with its statement in the airworthiness directive. The statement read "Ice accretion characteristics of the normal diameter droplets, as specified in the FAR, were entirely satisfactory. This confirms that model ATR-42 and ATR-72 series airplanes fully comply with performance requirements relating to the icing envelope specified in Part 25 of the FAR for certification of these airplanes."

The FAA proceeded contrary to the normal rule-making process by eliminating any chance of public comments from interested parties:
As described previously, the existing AD [airworthiness directive] imposes severe restrictions on many airplanes operated in air transportation, causing significant cost to the operators, widespread disruption of passenger travel, and an undermining of public confidence in the safety of the airplane. Based on the results of recent flight tests, if certain procedures and training are accomplished, these restrictions are unnecessary to ensure an acceptable level of public safety. Therefore, the FAA has determined that it would be contrary to the public interest to continue these restrictions during the period of time that would be necessary to issue a Notice To Solicit Public Comment on this action and to issue a final rule. Such notice is therefore also impracticable.

The FAA press release also announced the requirement for pilots to receive classroom training in the new procedures before operating an ATR in known or forecast ice; the training had not yet been developed, but ATR was to handle that with the airlines. Six items were covered as new issues in the press release in addition to the required training: no operation in freezing rain or drizzle; monitoring for the trapezoidal ice formation; no flaps when holding in icing conditions; no autopilot in freezing rain or freezing drizzle; immediately exiting freezing rain or freezing drizzle; and not retracting flaps if they are extended in freezing rain or freezing drizzle. Each one, with the exception of the new visual cue of the trapezoidal ice formation, had already been covered in November's operating limitations.

The visual cue of the side window trapezoidal icing indicating the presence of freezing rain or drizzle was known to Jeff Gagliano and Orlando Aguiar, as was the long-standing Simmons Airlines doctrine of not operating in freezing rain. I know the paranoia both pilots had for the aircraft in ice, and especially in freezing precipitation, and it is unlikely they saw the side window visual cue. We did not see it in the morning during our flights through the same weather, and I have not met any other ATR pilot who encountered it that day.

The only thing new was the premature indication of the requirement to install a "modification that precludes the formation of hazardous ice accumulation during flight in freezing rain or freezing drizzle conditions." The boot would do it, if approved, which it was. The confusing part of the wording is the suggestion that the ATRs might now be able to tolerate freezing drizzle or freezing rain, unlike every other aircraft operating anywhere in the world. In addition, the proposal indicated that after June 1, 1995, these rules would no longer be effective, and the ban on flying in known or forecast icing would again be enforced, unless the aircraft were modified with the new boot or other approved device. Apparently, the training and procedures were safe until June 1, 1995, but unsafe on June 2.

At the press conference on January 11, 1995, David Hinson stated, "Aviation is a science, and this decision represents a careful and methodical evaluation of the circumstances. Our decision was based on hard data gathered from many scientific tests. Since the Roselawn accident, we have taken prudent action each step of the way based on data available at the time." These words appeared again almost verbatim in a letter nine months later to 11-year-old Gina Gagliano in response to a plea for help she sent to President Clinton.

The airlines were happy. ATR was cautiously optimistic. The flying public and politicians were convinced by the spectacle, but the ATR aircraft were once again in harm's way. It was not long before the problems began again.

American Eagle had moved its ATR fleet south for the winter, so the lifting of the ban had a minimal effect on Eagle operations, but it did help the already reeling airline. Passenger confidence was down, and so were bookings on the Eagle aircraft. Unhappy Saab 340 pilots from southern routes brought in to fly the ATR routes out of Chicago and New York had rebelled during their first encounter with the mid-western winter weather. The airline essentially closed down the Eagle operation at Chicago O'Hare during the Christmas holidays. In Miami and Los Angeles, ATR crews from American Eagle's Chicago and Dallas hubs were flying under the new rules.

The training was based on ATR's publications, which the crews referred to as "the new car brochures." Dan Rodts called me shortly after his training, and he was irritated. "I feel like I just bought a new car, and they were still trying to sell me on how good it was. The glitzy brochure doesn't tell us anything new, and this side window stuff we knew years ago." The response was the same from other pilots, and most had already made up their minds about the ATR. The Simmons pilots who were used to the harsh realities of the northern winters were going to be cautious. The pilots from southern routes, who rarely saw much ice, were the ones to worry about, because they did not believe the stories about problems.

The first icing event to occur after the lifting of the ban caught another Chicago-based Simmons crew on temporary assignment to Los Angeles by surprise. Even with all of the operational regulations in effect and the American Eagle ATR fleet and pilots relocated to warm weather bases in Miami, San Juan, and Los Angeles until the spring of 1995, the crews still encountered icing conditions. Flights over mountains between warm-weather destinations, such as Las Vegas and Los Angeles, during late winter months found the aircraft routinely picking up large accumulations of structural icing at the required high cruise altitudes.

The operational locations had been changed by the airline for safety and confidence. Special weather collection and dissemination systems had been placed into effect to enable pilots and dispatchers to plan or reroute flights to keep aircraft clear of problem icing areas. The FAA rules had been reviewed and modified to preclude any more ice-induced control problems, allowing virtually unlimited operation of the airliners until the larger boots could be installed. The pilots had been trained to identify and leave dangerous weather conditions should they be inadvertently encountered. Despite these steps, the aircraft were still proving to be sensitive in some types of ice.

On March 4, 1995, one day after the conclusion of the Indianapolis NTSB hearings on the Flight 4184 accident, an American Eagle ATR-42, Flight 3947, piloted by Captain Roy Carmen and First Officer Pat Bauder and operated by Simmons Airlines, experienced an attention-grabbing, uncommanded roll event while approaching the Las Vegas airport. The aircraft was descending at an indicated airspeed of 230 knots, well above the prescribed minimum speed in icing conditions, and had accumulated about three-fourths of an inch of ice.

While the first officer was hand-flying the aircraft, he reported a buzzing in the controls from the ailerons and higher-than-normal control forces. The crew reported a roll excursion, according to a confidential internal Eagle memo issued to all flight crews by Simmons Airlines ATR fleet manager Al Rumph, who added a caveat to the bottom of the crew notification: "It should be noted that the manufacturer makes the following claim: Aileron forces are somewhat increased when ice accretion develops." The NTSB investigated the incident, as did the manufacturer and the airline. However, the public was never informed of the event by the airline or other official sources. Side window trapezoidal ice was not seen by the crew.

The ATRs moved back north in April 1995 and resumed normal operations out of Chicago O'Hare by the middle of June. American Eagle ATRs also resumed service out of New York to the northeastern United States. The entire fleet had been modified with the new, larger wing boots by the FAA's June 1, 1995, deadline.

The June 1 date was selected because the summer months in the northern climates are not considered to be aircraft icing weather because of the warm surface temperatures. The dangerous icing season around the Great Lakes is from October through April, but aircraft icing also takes place during the other months of the year. Sometimes it can be even more dangerous in July than in December because of the great amounts of moisture contained in the warmer air.

The atmosphere cools with altitude at the rate of 2 degrees Centigrade, or 3.6 degrees Fahrenheit, per 1000 feet above the earth. So on an 80-degree Fahrenheit day on the ground at Chicago O'Hare, the temperature of the air at 13,000 feet is 32 degrees Fahrenheit. This rate is a scientific average and the average rarely exists in nature, so pilots can expect to encounter prime structural icing temperatures between 10,000 and 15,000 feet. The rising and mixing of air that occurs during summer days, often identified by vertical, puffy cumulus clouds and evidenced by the air-pocket bounces an airline passenger experiences; high humidities; and the capacity of the warmer air to hold more moisture can induce significant amounts of structural icing. In thunderstorms, vast amounts of water can be transported to above freezing levels so fast that the water becomes supercooled. Supercooled water freezes on impact with an aircraft, creating dangerous clear ice or, at higher levels, rough rime icing or a combination of the two types.

Structural icing during the summer occurred shortly after the ATRs returned to Chicago. A fully trained Simmons Airlines/American Eagle flight crew was operating its ATR-72 aircraft between Toledo and Chicago on June 26, 1995, when the pilots encountered cumulus clouds at cruise level of 16,000 feet. The aircraft was equipped with the new, larger deicing boots on the wing, and the aircraft was indicating 200 knots of airspeed with the flaps up in normal cruise flight. The temperature was below the minimum for flight into cloud or precipitation without the deicing equipment operating, so the crew turned on all required systems. On entering the clouds, the aircraft began to accumulate ice, so Captain Chuck Schooler requested the wing deicing boots be activated to break the accumulation and remove it from the airplane. The crew watched for the side window icing as they had been trained. There is no mention of any other type of indicator of freezing rain or freezing drizzle or any training to detect other types of ice that have a negative effect on the ATR performance.

The aircraft had been flying in and out of clouds for a short time when one of the pilots noticed a ridge of ice above and below the wing that was not being removed by the larger deice boots because the ridge was beyond the new boots. Even with the new, larger boots and a crew that had completed the mandated FAA training on safe operation in ice, the airplane safety equipment clearly was not performing adequately in this weather situation. No side window icing was visible, and there was no other visual cue for which they had been trained to watch. Quickly and carefully, the crew descended to warmer temperatures, and the ice ridge melted. This option to descend to warmer temperatures is not available to crews in the winter, when the freezing level is on the ground.

I first met Anne McNamara, senior vice president and chief counsel at AMR Corporation, during the airline's investigations into my activities with the media, with a second meeting just before I was fired for speaking out. (Obviously, the airline did not officially terminate my employment for talking to the press about important safety issues, but it assembled several minor charges dating back to 1991 during its seven-month investigation before letting me go in June 1995.)

The first meeting centered around what was known about the ATR performance history in icing, and even though I should have considered Anne McNamara an enemy at that point, the need to have the help of senior AMR management transcended petty personnel issues. Anne indicated she had been completely in the dark regarding the past ATR problems but that AMR would do its own investigation. Feeling that she was honestly expressing her concern, I said, "Ask your pilots. Get their opinions and give them immunity from any type of retaliation so they will be comfortable in speaking freely." She never did.

When friends from the airline brought the details of the June 26th flight to my attention about a month after the incident, it was obvious that problems existed with the new equipment and procedures with the ATR, so I contacted Anne to ensure the flow of information had not been intercepted again. It was important that AMR Corporation knew of the holes in the system so they could act. Several days later, Anne confirmed the existence of the event and stated that AMR was working with the FAA on the investigation of the circumstances of the Chicago-to-Toledo flight.

My next contact was Jerry Barron, the Simmons Airlines principal operations inspector in Dallas. Since Jerry had previously supported the FAA position on the ATR and icing, after my Marquette incident, I was surprised when he provided the name of the person at the Seattle Aircraft Certification Office to whom he had forwarded the reports. When asked about his level of concern over the incident, Jerry had no comment. I called Seattle.

Gary Lium had testified at the NTSB hearings in Indianapolis regarding the certification issues with the ATR. He stated he was not involved in the original certification of the aircraft, but a document later surfaced with his name as the second addressee. This October 10, 1985, memorandum from Frank vanLeynseele regarding his precertification concerns over the performance of the ATR-42 in icing had been long forgotten.

After I explained why I had contacted him, Gary attempted to defer to the Public Affairs Office, but persistence gained some information. Yes, he had the reports regarding the June incident. Grasping at straws, I told him that I had heard of others as well, and, to my surprise, he confirmed that he had received four or five reports of similar occurrences since June 1, when the ban was lifted.

"Gary, what do you think about the building of ice in a ridge behind the new boot when it was clearly stated by Anthony Broderick this would never occur again? And how disturbing is it to you that the visual cue of side window icing was not a part of these reports?"

"We are in the process of determining how we are going to respond to those questions," he said.

Two months later, the FAA had determined its response, according to a reporter at the *Chicago Tribune*, who received this response in answer to a query: "FAA cannot verify any incident relating to new equipment and procedures for ATR. FAA has been notified of the reported event and the agency is looking into the matter." Another month went by. The final word given to the *New York Times* was along the lines that the agency had concluded the pilots did not see what they reportedly saw. The case was closed by the FAA without any further known action. The pilots involved were irate.

ATR conducted icing tests over the desert in California in early 1995. KC-135 icing tanker tests were flown with the 452nd Flight Test Squadron of the U.S. Air Force at Edwards Air Force Base. ATR contracted with the Air Force to perform the tests; the resulting documentation was turned over to ATR or ATR and the FAA. In the tests, an ATR-72 was flown behind a tanker fitted with a special head on its refueling boom to trail water mist dyed yellow, in effect creating an artificial cloud 9 feet in diameter. In the dry desert air, water evaporates rapidly, and it is important to position the aircraft at a proper distance from the tanker to obtain the desired results. The yellow dye allowed the ice to be easily seen as it accreted on the aircraft; the dye used had a minimal effect on the properties of the water vapor. The Air Force has used the tanker for years and understands that the program has limitations, which do not come anywhere near duplicating a real-world environment. Two separate flight tests were undertaken, one in an unmodified ATR-72 and the other after the NTSB hearings with an ATR with the enlarged wing boot.

The 9-foot-diameter cloud was first measured for droplet size and distribution by a Learjet flying behind the tanker carrying special instruments. Then the ATR moved into position, and the carefully planned flight profile was flown. Unlike the testing done at Bascombe-Downs in 1988, the tail was neglected during the flights, and the ice cloud dispersed so rapidly in the dry air that the stream was not even visible by the time it reached the tail of the ATR. Icing only a small section of the wing meant the aircraft would not behave the same way as an ATR with ice on the wing and tail, so the engineers had to extrapolate the data to determine the anticipated effects on the entire craft.

The ATR test pilots studied the control responses exhibited by the accident aircraft as well as the wealth of information on the past incident ATRs. The test pilots knew what to expect, when to expect it, and how to adequately respond to the anticipated control forces. Ironically, the test pilots also planned a response time lag between 1.5 and 2.0 seconds before they initiated recovery. The test's designers said this time lag was the time that the crew of Flight 4184 had to respond; it also corresponded to the reversal of the controls recorded in the accident flight and numerous incident data tracks.

Almost on cue, the ATR departed controlled flight when the ailerons deflected on their own. The test pilots left the tiny cloud and recovered the aircraft with smooth control inputs. ATR carefully protected the confidentiality of the tests, and even six months later, the Air Force would not discuss the matter. Inside sources at the Air Force, however, confirmed that at least one test "got a little dicey." The press releases indicated total success and confirmation of all anticipated data.

One of the photographs taken after the ice test showed a severely iced ATR nose section, and ATR touted the still photograph and videotape of the yellow aircraft nose as an example of the extreme nature of the testing. One ATR pilot correctly analyzed the photographs by saying, "Big deal; the nose doesn't fly the airplane, and the wings and tail were completely clean."

Although the larger boots provided at least a minimal increase in the protection of the wing, the manufacturer still refused to address the horizontal tail by installing a greater percent of chord protection for this critical surface.

The training at other airlines caused some pilots to conclude that the problems were more with the flight crews than the aircraft. Pilots from Miami who had never even been in an ATR chimed in with opinions in a *Professional Pilot* magazine survey during the late summer of 1995. Several Simmons pilots, however, who have thousands of hours in the aircraft operating in the northern United States, exhibited an informed opinion hammered out of experience. The survey might have shown the lack of information understood by the writers of the letters in response to the questions. The magazine article reported 82 positive, 33 negative, and 11 neutral responses to a question about how pilots judged the ATRs overall. As with the feedback from some of the pilots at Simmons when talking with me, many responding pilots parroted the ATR company line, which is a dangerous situation.

A professional polling organization painstakingly designed a survey under the sanction of the FAA, which provided a list of ATR pilots in the United States. The goal of the survey was to develop additional information on past incidents that might have gone unnoticed as well as to grade the training received in dealing with various ice-related anomalies. After taking several months to develop the survey to ensure fairness, an initial group of 50 surveys were sent to pilots, many of whom worked for American Eagle. Some took the questionnaire to the company management, who relayed the information to ATR. The FAA was contacted. Significant pressure was brought to bear on the individual who prepared the poll, and he was even threatened with legal action. The survey, which might have collected valuable information, was squashed.

During September 1995, word reached me from four separate sources across the country that the FAA was discussing the possibility of decertifying the ATR aircraft in the United States. Even pilots taken to the American Eagle headquarters in Dallas for interviews were told that on September 21, Ralph Richardi, the new president of Simmons Airlines, and Lance McDonald, Simmons' vice-president of flight operations, were in Washington, along with senior management from other airlines operating the ATR, to discuss the issue.

A source in the NTSB responded to my inquiry about the meetings and the possibility of decertification by saying "That will never happen." When asked to check on whether the discussions were, in fact, occurring, he agreed to verify the information but repeated, "They will never let that happen."

The former principal operations inspector for Simmons Airlines, Anson Gray, told Dan Rodts that the airline had been lobbying against the decertification before Simmons was moved to the Dallas office of the FAA and that they continued to fight for the issue. An American Eagle pilot who dealt with the ATR safety issues in his position claimed that the discussion had reached his level after meetings with management about new operating procedures for the aircraft during icing operations. A senior AMR Corporation official reported the airline would be "foolish to purchase any more ATRs" because numerous issues still needed to be worked out. AMR lost money during the disrupted service in the winter of 1994-1995 and had presented a demand to ATR to cover the losses.

Anne McNamara replied to my inquiry regarding the meetings by stating that the French held a trump card in a controversy over the Boeing 737 jetliner. Because the crash of USAir Flight 427 in September 1994 in Pittsburgh was still officially unsolved by the NTSB, the FAA was pressured by the possibility of retaliation. If the agency decertified the ATR, the French DGAC would take similar action against the Boeing jet. The FAA not only received pressure from politicians supporting the French position but also from those representing thousands of jobs that could be affected at Boeing, a U.S. manufacturer. The Boeing 737 is the most successful transport aircraft in history, and it would not serve the public confidence to have a U.S. ally acting against it on a safety issue.

AMR was doing a few positive things. It was working with the FAA and others to develop a database on icing so they could track all incidents. The company mandated ATR winter operations training for its pilots and set up a safety hotline. To the credit of AMR, it took some action in the interest of safety.

The FAA lobbied the independent NTSB and chief investigator Gregory Feith on behalf of ATR. On October 17, 1995, Feith was summoned to a meeting with the FAA to convince him to soften the language against the aircraft in the final report. It also was the date that Greg informed Cindy Gagliano that the hearings that would make public the conclusions of the board would be delayed until spring of 1996 as the French moved to protest the report again. Even though the cause of the accident had been nailed within two days of the accident, it was more than 16 months before the families could move to put things behind them. They saw the first anniversary of the crash come and go while the NTSB wrestled with politics.

One expert floated a theory that one reason the ATR-72 wing is so friendly to the formation of structural icing has to do with its composite construction and the efficiency level of heat transfer. It is possible, he suggested, that because the wing is mainly a composite material, it can be cooled at high altitude and will retain that "cold-soaked state" much better than an aluminum wing, which readily adapts its operating temperature to the ambient conditions because of its excellent heat-transfer rates. The cold-soaked wing could possibly attract moisture that is above freezing and lower its temperature to the point at which the water would freeze on the wing. Although composite materials are being used frequently in the construction of many new aircraft in addition to the ATR, it is difficult to find another turboprop airliner with all-composite wings.

Back in 1988, when the U.S. agencies were maintaining distance from the Italian investigation in an effort to avoid offending either the French or Italian officials, C.O. Miller, a 40-year veteran of aircraft accident investigation, worked with the Italian airline on the causal factors surrounding ATI Flight 460. After the crash of Flight 4184, a number of similarities with the Roselawn crash came to light, and he contacted the NTSB with an offer of help.

Miller had briefed both the FAA and NTSB on the Italian accident in 1988 and made three separate inquiries of the board after the crash of Flight 4184. In early November 1994, his letter to Chairman Jim Hall was greeted with a thanks-but-no-thanks telephone call from Ron Battocchi, an NTSB staffer, indicating the staff was aware of the findings of the Italian accident. He did not get a response directly from the NTSB to his second letter of June 1995.

Miller raised several key points with public NTSB information from the investigation and correlated it with his knowledge of the Como, Italy, crash. Conflicting information over the exact status of the elevator uncoupling mechanism in the Roselawn accident was found in the reports. The first inspection revealed that each side of the elevator moved together, but subsequent investigation showed they moved independently, meaning there might have been a control circuit break before impact. ATR told the board there had never been an inflight disconnect of the uncoupling mechanism, although they used the phrase "during a revenue flight." The Como case findings showed that ATI Flight 460 had to deal with each pilot flying the opposite sides of the tail, and this situation had been duplicated by other pilots from the same airline.

Miller discussed an *Air Safety Week* magazine report from May 22, 1995, that stated ATR had still not been able to "recreate the accident" and that "no combination of factors have been completely consistent with the Roselawn accident." This fact was confirmed by ALPA pilots relaying information that ATR had attempted hundreds of different scenarios without success in an attempt to define the accident. The disconcerting part of the manufacturer's inability to duplicate the behavior of the aircraft in the accident is that apparently no attempt was made to investigate any possible role the tail might have played in the crash. Miller stated, "More to the point, however, are discussions I had with stability and control experts at NASA Lewis Laboratory and a well-known icing expert in Sweden during the ATI investigation. They were quite firm in their belief that lateral/directional control can be influenced markedly by icing of the empennage based on flight tests undertaken in Cleveland seven or eight years ago and previous accident investigations. Of course, this is consistent with basic aerodynamics even as I remember teaching it years ago."

Miller laid several questions at the feet of the board. "Is there something we possibly missed? What about any loose ends? Have we really examined all of the avenues for preventing this kind of accident in the future?"

A couple of weeks later, Miller received a phone call that cannot be linked directly to the NTSB. A soon-to-retire board member, who was part of the investigation at the hearings in Indianapolis, called and reportedly asked Miller to "stop stirring the pot on this ATR issue." Although no one could ever trace the connection to the NTSB, Miller believed he was acting on a board request.

ALPA published a three-part report on icing that looked at the certification standards, the hard choices of risk versus reward, and operations. The final issue provided the most informative insight and also showed a Saab 340 aircraft flying in conditions exceeding the certification standards. Although the aircraft accumulated ice, no buildup of ice occurred behind the wing boots. ALPA's pilot magazine articles noted the commitment of the FAA to more research, the contribution of the union to the controversial solution, and the need to examine operations beyond those specified in Appendix C of the FAR regulations.

Experts have been warning the FAA since the mid-1970s that the Appendix C envelope developed in the late 1940s was obsolete, and numerous reports written and stored since that time have had no impact. The government's solution apparently is to spend more tax money on research that has already been completed.

A document translated into English from the file on the Italian accident stated the situation with regard to the ATR and operating conditions. "It came, in addition out of the accident—and from other previous incidents—that the constructors had to consider, beyond certification requirements, the difficulties that could arise in practice from the actual experience and knowledge of some of the pilots employed by regional airlines and the greater exposure of turboprop aircraft to icing conditions than jets." ATR aircraft and all turboprops operating at busy airports such as Chicago O'Hare have a more difficult task than the jets. ATRs operate in the weather. Jets are given unrestricted climbs to the safer, higher altitudes, whereas ATRs fly at lower altitudes, in rough weather, turbulence, and ice for lengths of time much greater than the eight minutes of ice exposure it took to bring down Flight 4184. Even though the ATR is as capable as a jet at less than 10,000 feet, it is exposed to greater perils of aviation than the jet.

It all comes down to money. When the FAA recently caught ATR doctoring official FAA documents on a propeller to indicate that a used mechanism was new, the company claimed it was an oversight and that it "did not do business that way." History has proven otherwise.

Years ago, Simmons Airlines had an ATR-42 that was sometimes reported by crews as weaving through the air. Maintenance never figured out the problem, and finally it was decided to send the aircraft back to Toulouse to have the rudder replaced. Deicing fluid, it was learned, had saturated some of the honeycomb material in the rudder, causing an imbalance that caused it to flutter in flight, producing the wagging motion of the nose of the aircraft. ATR replaced the faulty component and the aircraft returned to service, never to exhibit the tendency again.

Several months later, another airline in a different part of the world reported an ATR-42 that weaved. Investigation revealed that the manufacturer had taken the Simmons rudder and swapped it with the rudder of another ATR on the assembly line.

13

The past promises of the FAA

Transportation Secretary Federico Peña apparently buys into the philosophy of style over substance. He articulated his belief during an interview published in *USA Today* on December 22, 1994: "It's important not to focus so much on the statistics, but (on people's) perceptions." Apparently, for officials looking to assuage public fears, the pronouncement supersedes the accomplishment, and because Secretary Peña oversees both the FAA and NTSB, his doctrine probably is carried out in his agencies as well.

The Federal Aviation Agency was established in 1958 to replace the Civil Aeronautics Administration and became the precursor of the modern-day Federal Aviation Administration, which in 1966 became part of the Department of Transportation. The FAA was responsible for pilot licensing, aircraft and airline certification, air traffic control, and regulation development and enforcement. In 1967, the National Transportation Safety Board assumed responsibility to investigate all aviation-related accidents. Legislators provided the FAA with an interesting dichotomy in its charter that required the agency to both regulate and promote aviation in the United States. That dual role often causes the FAA to find itself on both sides of a safety issue, with its own management being the arbiters.

To help with decisions, the FAA is required to prepare a cost-benefit analysis before it imposes costly changes on the industry it regulates. In calculating cost, the FAA may ask the airlines, for example, how much it would cost to institute additional training, and the airlines respond with an estimate. The FAA then calculates how many lives would be saved by the imposition of the new regulation and multiplies that number by $2.5 million, which, according to the FAA, is the value of a human life. On paper, the system may make sense, but with well-funded lobbying efforts mounted by the Air Transport Association, which represents the airlines, the case is rarely as simple as the figures indicate.

Throughout its history, the FAA has been reputed to use "tombstone regulation," implying the agency would be unwilling to decide on a safety issue until the tombstone count was sufficiently high that public pressure precluded it from moving in any other direction. The list of names of people who paid for FAA inaction is lengthy, and the list grew by more than 250 names in 1994. The direction the FAA takes does not seem to matter if the presidential administration is Republican or Democrat or if the FAA chief is perceived as propassenger or proairline. Management of the FAA does not depend on pontifications of a zero-accident goal. The accident rate varies year to year, but the causes of those accidents usually fall into several well-known and long-understood categories.

The FAA is a bloated bureaucracy—on that point even the agency itself agrees—and industry insiders who deal with the FAA on a routine basis describe it as a poorly managed agency with cumbersome solicitation procedures that slow acquisition of new technology. By the time the purchasing rules of the agency can be complied with, the acquired technology is often out of date before it becomes operational. In some cases, equipment promised for safety enhancement during intricately choreographed and widely publicized press conferences sits idle in warehouses because of the agency's failure to properly manage the projects or even enact or enforce the provisions of eminent domain.

The media showed startling images of the charred and smoking remains of a Delta Airlines L-1011 widebody jetliner that crashed while attempting a landing during a thunderstorm at Dallas-Fort Worth Airport on August 2, 1985. Of the passengers, 32 survived; 132 aboard the aircraft and one motorist on the ground died. The jetliner was less than one minute from landing when it flew into a rainshaft and encountered a severe shift of wind direction and speed, causing a descent that the crew was unable to arrest before the airplane struck the ground, skipped over a highway, slammed into a water storage tank, broke apart, and burst into flames.

The investigation revealed that the 400,000-pound aircraft encountered a windshear phenomenon called a microburst. Officials began to use this name after the downing of a Pan Am Boeing 727 in New Orleans in 1982. This newly discovered phenomenon was different from conventional windshear, which is a sudden or drastic change in wind speed or direction. A microburst is defined as a severe, short-lived, localized downdraft normally associated with a powerful storm cell. The itinerant nature of a microburst, existing for 10 to 15 minutes with peak intensity lasting about two minutes, and its small size of about 2 miles in diameter, belie the danger of the phenomenon, which can generate wind speeds of more than 100 miles per hour.

The Delta crew might have been lulled into a false sense of security because it could see the airport beyond the rain shaft and because a small business jet had just passed through the same area, reporting only moderate rain. On entering the rainshaft, the Delta crew failed to recognize the windshear as the airspeed of the airliner started to increase and the aircraft floated above the electronic glideslope, and the pilots reduced power to correct the flight path deviation. That decision was fatal.

When the L-1011 flew into the tailwind portion of the microburst, the airspeed quickly dissipated, the application of power could not stop the rate at which the aircraft was sinking, and the aircraft hit the ground. A tower controller had issued an urgent order for the flight to go around when he saw the big jet emerge from the rainstorm at only 50 to 100 feet above the ground. Two or three seconds later, he saw the left wing and nose of the plane hit the ground, and the aircraft burst into flames. The jet hit the ground with a level attitude, and miraculously, survivors crawled from the burning and smoldering wreck.

The Dallas-Fort Worth Airport was equipped with a first-generation low-level windshear advisory system that consisted of a wind direction and speed sensor at the center of the airport and five locations around the perimeter of the airport. The six sensors fed data into a central computer that compared data from each sensor against the data from the center of the airport and issued warnings to controllers when differences in speed or direction exceeded set parameters. The systems had been in operation at airports in the United States since 1979, but it had been criticized for being late in generating warnings and ineffective in advising controllers of the existence of the dangerous microbursts that are small enough to exist between sensor locations and cease to exist by the time a warning is issued.

Scientists had been working on advanced warning systems to pinpoint microbursts for years before the Delta jet crashed in 1985, but it was six years before the new Doppler radar system was announced. In the interim, an expanded system that increased the number of sensors for low-level windshear advisory systems from six to 16 and moved them further from the airport perimeter to provide wider coverage was installed. Along with additional sensors, new computer software targeted at the special flow characteristics of microbursts was announced, but it was not installed until 1989. Even though it was only expanding the coverage and software using current technology, the FAA was unable to keep the systems it did have operating in working condition. The sensors were prone to damage from lightning, often leaving controllers and pilots without the critical information when they needed it most, during thunderstorms.

The new sensors, when available, would be placed on public lands within about 1.5 miles of the airport boundary along extended centerlines of the runways. Should public property not be available where it was needed to place a sensor, Greg Goff, the nationwide manager of the low-level windshear advisory systems for the FAA, stated the federal government would move to condemn and use private land. The expanded systems and new software would theoretically double the effectiveness of the system, but that would still only mean a prediction rate of 80 percent for windshear.

On January 6, 1991, the FAA announced a new Doppler radar system designed to warn pilots of potentially deadly windshear. The hope was that the new system could cut the rate of accidents caused by windshear from one or two per year to one every 20 years. With the new system, controllers would receive a 10- to 20-minute advance warning of the development of a microburst. The systems would be operating by 1993 at airports around the country.

Doppler radar measures velocity and distance by bouncing microwaves off objects such as rain, snow, or even leaves or insects that are moved by the wind, and translates that information into digital images. The designer of the system for the FAA was Raytheon, a defense contractor that had a commitment to supply 47 systems at a cost of $190 million. Once the first 47 sites were in operation, the government had options for another 55 systems.

Although trained meteorologists working with a similar Doppler radar system in 1984 had proven that constant monitoring of the data allowed them to predict a microburst several minutes before it occurred, the FAA announced that the new system, installed in an operating air traffic control facility, was able to give a 10- to 20-minute advance warning. Doppler radar, which had been in use for nearly 40 years, was developed by the military for guided missile control, but the recent advancements in digital electronics allowed the equipment to be used in civilian applications. Less than 60 days after this announcement by the FAA, United Flight 585 crashed during an approach to Colorado Springs, Colorado, killing everyone on board. This crash remains one of the three unsolved airline accidents in NTSB history, but all indicators point to windshear.

The installation of the Doppler systems by 1993 was to be a welcome safety addition to the system. On a stormy summer day in Charlotte, North Carolina, USAir Flight 1016 approached the airport. The

DC-9 airliner slammed into the ground in a residential area, killing 37 of the 52 people on board. The crew saw a large shaft of rain near the airport and had determined a route of escape if the ride got bad. Several jets elected to wait until the storms passed before taking off. At 6:23 p.m. the crash call came; the date was July 2, 1994.

In Senate testimony during January 1995, FAA administrator David Hinson stated that only two Doppler systems were in operation in all of the United States. The other systems were in warehouses while the FAA negotiated with landowners about the placement of installations. Ten miles from the impact location in Charlotte, a selected Doppler radar site waited for equipment that would arrive too late for those on the DC-9. The government can condemn property to place a highway through farmlands but cannot, with all its power and financial resources, secure the rights to install radar towers on someone's property. Television stations in small markets would have the Doppler radar systems for their on-air meteorologists to predict rainfall long before the United States government could place them to protect air transport.

The public believed the Doppler systems were in use because the FAA had said they would be three years earlier. Most people had forgotten about the announcements after the last accidents, and even the vision of the smoldering, broken Delta L-1011 was a distant memory. Nearly 10 years after the press conference proclaiming the new system and FAA promises, people are still dying because of bureaucratic ineptitude.

Recent interviews with top FAA officials revealed the position of the agency on another critical safety issue—fraudulent aircraft parts finding their way into airliners. Individuals in aviation understand the increased costs associated with specially manufactured and certified parts labeled "aircraft." Even though that spark plug, oil filter, or hydraulic line appears identical to the off-the-shelf counterpart for automobiles, and might be, with the label "aircraft," the price skyrockets.

For-profit operations, such as major airlines, have had a problem in recent years with bogus aircraft parts; that practice has grown to a billion-dollar-per-year business. International organizations are concerned about the growing scope of the racket, which has become more lucrative than the drug trade. Several crashes, some first blamed on pilot error or structural failure, are beginning to plague aging fleets of airliners and have been traced to fraudulent or undocumented aircraft parts.

The FAA has a program that is designed to protect parts purchasers from fraud. It documents the history of each aircraft part from manufacture to scrap heap. All critical parts are given a serial number, and each part placed on an aircraft is logged. When, for example, a starter is replaced, the old mechanism is removed and tagged for re-manufacture at a licensed shop, and the new part is put in place with log entries to document the operation for FAA inspectors or future owners of the aircraft. Log books remain with the airplane from the date it is manufactured to the date it is retired, and the attempt to sell an airplane or major component that has a nonexistent or a reissued log is highly suspicious.

A recent television program detailed the case of a British Airways Boeing 747 that was destroyed in Kuwait in 1991 during the conflict with Iraq. Fire destroyed the aircraft, and insurance companies paid the owner for the full loss and hired a reputable salvage company to remove and destroy the hulk. The underwriters insisted on videotaping the destruction of some of the more recyclable parts to ensure compliance with the contract. Months later, even though the salvage company had supplied a videotape of a workman cutting the turbine blades of a jet engine with an oxyacetylene torch, the fire-damaged engine and auxiliary power unit showed up, complete with documentation, for sale by a certified parts reseller. Other parts from the scrapped jumbo jet turned up around the world.

Mary Schiavo is inspector general for the Department of Transportation. She has completed an in-depth study of the bogus-parts problem and classifies the issue as a ticking time bomb. The trafficking is global, and she has only 75 investigators to track down fraudulent activity. A single jetliner contains hundreds of thousands of parts that come from manufacturers around the world. The difficulty in tracking parts from scrapped airliners or damaged used parts is obvious. Some reputable shops have unscrupulous employees who make additional parts during a production run and sell the extra, undocumentable items for a huge discount to wholesalers who provide the bogus certifications. The chance of being caught is low, and new operations spring up all the time.

Even the FAA is not beyond being duped by sophisticated parts cheats. A survey of more than $32 million in aircraft parts in storage for use in the FAA fleet of aircraft found that nearly one-third of the inventory did not meet the agency's standards. Over $3 million worth of parts were scheduled to be destroyed, further wasting taxpayer funds. However, a statement by FAA assistant administrator Anthony Broderick about the problem invokes incredulity. "The FAA is not required to follow its own rules," he said during an interview on a PBS special that aired in September 1995.

For years, air traffic controllers have complained about the archaic and unreliable radar computers in use for separating aircraft. During the summer of 1995, in an act of frustration, Lauren McCormak, a veteran air traffic control specialist with 11 years of experience working at the Aurora radar facility, publicly resigned.

Aurora, Illinois, is the home of Chicago Center, which controls 120,000 square miles of airspace above five Midwestern states, including aircraft inbound and outbound from Chicago O'Hare. The Chicago Air Route Traffic Control Center is the busiest facility in the United States and is responsible for the safe separation of 9500 flights each day. Lauren McCormak decided to resign because, in essence, he believed that it is "only a matter of time before a disaster occurs" as a result of the antiquated equipment controllers use, and he did not want to be there when it happened. McCormak went public with his concerns to educate the public about the risks and move legislators to accelerate the pace of FAA acquisition of new equipment.

Less than one month after McCormak's announcement, Chicago Center in Aurora suffered its third, fourth, and fifth main computer system failures in less than one year; all failures occurred during one week in July. The sixth outage for 1995 occurred in mid-September. Fortunately, the less-capable backup system worked, and according to the FAA spokesperson, "There's certainly no danger that I can see this morning."

A spokesperson for the air traffic controllers' union was quoted in the *Chicago Tribune* with a different view of the situation. "If [a computer breakdown] happens when it's busy, then it's panic city. We've been extremely lucky."

The FAA reports that the new computer system, which will replace computers that date back to the 1960s, will not be available until 1997 overall and 1998 in Aurora. In 1987, the agency updated its computers, but they were not designed to handle the video displays in use. These new computers were linked electronically with the old equipment, and when one computer goes down, the whole system fails.

The company that builds the systems stated recently that the new systems could be operational within four to six months of the placement of an order by the FAA. At an estimated cost of $200 million for six center control facilities, "Money is clearly not the problem," stated one senior FAA official. The problem is the antiquated government procurement policies that block the movement of air traffic control system into the 21st century.

In 1981, President Ronald Reagan fired 11,000 air traffic controllers for engaging in an illegal strike over working conditions, pay, benefits, and the nagging problems with the overworked air traffic control system. Since then, air traffic has increased dramatically, yet the system remains virtually unchanged. By the mid-1980s, the number of air traffic controllers had not returned to prestrike levels, and the Air Transport Association, an organization that represents airline companies, proposed shifting the nation's air traffic control system to a new federal corporation. The move would replace the layers of FAA bureaucracy, its inflexible personnel policies, the intolerable civil-service regulations, and what was described as a paramilitary attitude at the top levels of the FAA. The idea was met with cautious optimism outside the agency. Inside, the FAA prepared for a fight to retain control over a major portion of its 50,000 employees.

In May 1988, another warning was issued about the outmoded equipment at both Chicago O'Hare and Chicago center, even though the equipment had been updated in 1987, and shortages of controllers and technicians, all of which contributed to decreased levels of safety. The Better Government Association studied the problems, and its conclusions agreed with reports filed by the General Accounting Office and the NTSB and indicated a number of red flags had appeared that indicated the need for urgent action.

The FAA refused to discuss the report but issued a conclusion in its statement that the system was safe and would be safe tomorrow. Controllers and technicians working at the center disagreed with the perception, not the statistics.

By late 1990, the number of daily air travelers topped 1.3 million, and the system strained to accommodate them. Three years earlier, 21 airports had suffered more than 20,000 hours of flight delays, but only two new airports were commissioned in the previous 20 years with only one more, near Denver, under construction. The FAA was facing overused airspace surrounding the major hubs that the airlines had set up in the 12 years since deregulation, and the agency seemed not to have a proper plan for the anticipated growth in traffic, which was expected to double in 12 years.

By 1991, outages plagued the system, and air traffic computers that rely on telephone lines to send data communications across the country revealed another vulnerable cog in the system. An AT&T switching center attempted an automatic switch from a commercial power source to a generator backup system. When that system failed, the center switched to batteries. After six hours, the batteries died. The telephone lines were out for seven hours even though the FAA had been assured the outage would not last longer than 30 minutes.

Congress criticized the outage as commonplace and not an exception, and the overseers demanded a plan to make the system failsafe so as to not place air traffic in jeopardy in the future.

In 1994, David Hinson announced budget cuts to the agency's $7-billion program to modernize the nation's air traffic control system and further indicated that only the air control towers at the largest airports would get the new computer systems. He also indicated that the new state-of-the-art system to expedite the flow of air traffic would be re-examined. Hinson issued assurances that the cutbacks would not jeopardize air safety or efficiency but would save the taxpayers hundreds of millions of dollars.

However, in 1995, with outages in the current air traffic control computer system mounting around the country, the FAA again faced pressure from Congress to accelerate the acquisition of new computer systems slated for 1997 or 1998. In August, Secretary Peña announced the FAA was complying with Congress, and the interim replacement system would be installed in the busiest of the air traffic control centers by 1997, with replacement computers slated for installation in 1999. The Secretary had come to the Chicago Center in Aurora shortly after the fifth computer outage of 1995 to give controllers a pep talk and assure them the government was doing all it could to solve the problem.

At the same time, the Clinton Administration was pushing to place the air traffic control system in the new federal corporation as suggested as far back as 1985. Because the FAA is required to comply with burdensome regulations, budget constraints, and procurement requirements, it was suggested a new government bureaucracy was needed to fix the old government bureaucracy. For 10 years Washington had toyed with the idea, sailing it time and again at press conferences as a possible solution after a problem became the focus of public attention.

The fifth failure at the Chicago Center was brushed off as a "16-second hiccup," wherein the computer sensed an error, dumped its data, and performed a restart. However, controllers did not know what was happening when their screens lost data on airliners with thousands of people on board. Even seasoned professionals can get rattled under pressure, and the uncertainty of the system adds to a high-stress profession. In addition, a jetliner traveling near Chicago, Dallas, New York, or Los Angeles and operating at the maximum speed of 250 knots at less than 10,000 feet travels more than 1.25 miles in 16 seconds. The chances for a close call or worse are high.

The stress level at high-traffic facilities makes it difficult for the FAA to enlist qualified controllers. A pay bonus of 20 percent was slated to be erased by Congress, and staff shortages required overworked controllers to put in large amounts of overtime. To increase staff levels to where they should be under FAA guidelines, Secretary Peña announced he was immediately sending 50 additional controllers to the understaffed Chicago Center. In late September 1995, with more than one month elapsed since the announcement, three national bids failed to provide enough volunteers to take positions in Aurora. Of the 42 controllers who had agreed to move, only nine had enough experience to move into positions immediately; the other 33 would need between 18 and 36 months of training to become qualified Chicago Center controllers. Three years will pass before the positions will be staffed by fully qualified personnel, one year after the new interim computer system is scheduled to arrive.

Public frustration seemed to peak within a week of an announcement by FAA chief David Hinson on September 7, 1995. He stated, "Statistics show that the air traffic control system nationwide has become increasingly reliable over the last seven or eight years with equipment operating properly over 99 percent of the time." He added a few words to assuage the increasing concern of the flying public by asserting "the air traffic control system is absolutely safe."

Five days later, on September 12, Chicago Center experienced its sixth failure of the year. Unlike the other occasions, this time two aircraft came close to disaster when an American Eagle ATR came within 700 feet vertically and 3.5 miles horizontally of a small private plane. News articles the next morning reported the incident and details of a similar incident nearly two decades earlier that led to the fiery crash of a PSA Boeing 727 jetliner at Ceritos, California. Fortunately, the traffic collision and avoidance system, a device that was recommended after the PSA accident but not installed on passenger aircraft until nearly 15 years later, was on board the ATR and warned the pilots. The pilots performed an evasive maneuver and prevented a collision. In an interview from Washington, David Hinson once again relied on style over substance to calm the public by stating the system is "Designed to be failsafe" and added "It is safe." Hinson surmised there was "no danger" in the incident. However, controllers saw it differently and issued a statement through Ron Downen, union vice president, claiming "safety has definitely been affected today." The conflict caused David Hinson to travel to Chicago two days later.

Both the FAA and the controllers agree that the system needs updating and staffing levels need to be increased. Controllers in high-density traffic centers should be well paid to compensate for the higher risk, and the bureaucracy needs to be reduced to make money available to design, purchase, and install the new equipment necessary to handle the increasing level of air traffic. In 1995, the problem is the same as it was in 1981.

While the FAA battled the air traffic control system and its controllers publicly for 15 years over change, another fight went on with the NTSB. On December 27, 1968, the NTSB investigated its first air transport category accident involving structural icing. The crash of an

Ozark DC-9 while attempting a takeoff from Sioux City, Iowa, was a result of pilot failure to check the ice accumulation and account for its aerodynamic and weight penalties before the takeoff, even though ground personnel had advised the crew of its presence. No safety recommendations came out of the investigation. In 1978, after another DC-9 accident, this time involving a TWA jetliner at Newark, New Jersey, on November 27, and prompted by the growing number of accidents and incidents related to structural icing, the NTSB undertook a special study on aircraft icing avoidance and protection. The study was adopted by the NTSB on September 9, 1981, and provided conclusions and recommendations.

Unfortunately, the NTSB has no regulatory authority and can only lobby the rule-making FAA for what the experts at the NTSB consider safety improvements. For years, arguments have raged behind closed doors between the FAA and NTSB about the adoption of the recommendations of the NTSB. In December 1994, David Hinson characterized the frequent disagreements between the two authorities as "a healthy tension."

On September 24, 1981, chairman James B. King of the NTSB wrote to FAA administrator J. Lynn Helms detailing a number of recommendations coming from the icing study. Under governmental rules of engagement, the FAA responded four days before Christmas. The NTSB had asked for closer monitoring and study of the characteristics of the atmosphere involved in aircraft icing to provide better forecasting using data instead of the subjective terminology in use at the time. The pilots then could make an informed decision about the limitations of their aircraft with the use of the weather forecast and its information on liquid water content, drop size distribution, and temperature of the clouds.

To develop data for each individual aircraft type, the NTSB urged a review of the Part 25 transport aircraft certification standards, which pertain to icing based on the latest knowledge of cloud physics and the characteristics of aircraft. Appendix C had been in use since 1948, and studies in the late 1970s had determined flaws in the certification guidelines, testing procedures, and real-world operations.

The icing tanker test used in post-Roselawn crash studies of the ATR had deficiencies known in 1981, as the report pointed out. The cloud produced by the tanker was not large enough to cover a whole aircraft, limiting the icing coverage to certain parts of the aircraft at any one time. Because the established parameters for the cloud were found at a fixed distance from the tanker, the liquid water content and drop size rapidly decreased as the distance between the tanker and the test aircraft increased. The study revealed in 1981 that limitations existed in tanker testing, and those limitations needed to be accounted for when accepting any data obtained from tanker testing. Because an aircraft functions as a single entity when encountering icing, icing a small portion of affected areas while the remainder of the aircraft is clear might not give an accurate representation of how the entire craft might respond to the accretion levels tested. Because the testing was unable to duplicate the naturally occurring conditions by covering enough of the aircraft's surface with ice, it is understandable that ATR was unable to recreate the behavior of the accident aircraft at Roselawn, even with more than 500 test attempts.

The FAA agreed with some recommendations and disagreed with others. By April 16, 1982, the chairmanship of the NTSB had been awarded to Jim Burnett, who responded to the FAA response. Of the four main NTSB suggestions, two had been classified by the NTSB as "Open—Unacceptable Action" and two were listed as "Open—Acceptable Action." The comments included requests to reconsider the proposals of the NTSB and pointed out the "operations in freezing rain, freezing drizzle, and mixed conditions occur often enough to warrant inclusion of such conditions in the certification criteria, especially considering their hazardous nature." Between the generation of the first letter to the FAA and the first response back, ATR was forming as a consortium and made application for a new aircraft, the ATR-42.

On June 7, 1982, the FAA again responded to Burnett and took the erroneous position on the certification issue that the current FAA standards "require an ice protection system which permits safe flight in maximum icing conditions," stating the rules do not allow for certification for less extreme conditions. Although the rule was designed to allow for limitations being placed on certain aircraft types to eliminate the subjective decisions made by pilots that might prove fatal to them and their passengers, it did not seek a variable standard for certification. The game of twisting the intended meaning had begun as a delay tactic, and later in the letter, the FAA changed its position, possibly allowing elective certification to higher standards such as operation in freezing rain for some aircraft. Although it implied the understanding of the severe danger posed by the phenomenon, the documentation stated that freezing rain was a rare occurrence and easily forecasted and therefore easily avoided.

More than one year passed before the NTSB again pushed the icing issues with the FAA. In an October 24, 1983, letter, the NTSB indicated that the report from the Federal Coordinator for Meteorological Services and Supporting Research on icing research and forecast and operational procedures was forthcoming. In the interim, the NTSB had performed a study on one year of icing accidents and found 28 percent involved encounters with freezing rain and rebutted the FAA by pointing out such a percentage was not a rare event. In addition, a previously "Open—Acceptable Action" item was reclassified by the NTSB as "Open—Unacceptable Action." The next day, on October 25, 1983, the FAA response was short but addressed only the modified status of the item pertaining to operation of aircraft in severe conditions, a regulatory loophole that still exists today even though no transport aircraft is certificated to operate in those conditions. The letter delayed any action pending the study done by the Committee of Aviation Services under the Federal Coordinator for Meteorological Services and Supporting Research but promised to keep the NTSB advised of relevant progress.

Sometime after J. Lynn Helms sent that letter and the next NTSB written contact on the issue on February 21, 1984, a new acting administrator, Michael Fenello, took over at the FAA. Fenello received Jim Burnett's letter, which advised the FAA that the NTSB accepted the delay in acting on the operating rules for airliners until after the study. The status was again changed to "Open—Acceptable Action."

The recommendations went dormant for nearly three years before the FAA responded on December 1, 1986. The year was the 50th anniversary of the existence of the FAA and its predecessor; Donald Engen was now the administrator of the FAA. The FAA made recommendations in an advisory circular, which is a voluntary paper carrying no regulatory power, and wanted the issue of expanded definition of the cloud physics closed. The language was strong enough to generate a negative response from Jim Burnett in March 1987. The NTSB closed the recommendation that desired formulation of an established, standardized "procedure for the certification of aircraft which will approximate as closely as possible the magnitudes of liquid water content, drop size distribution, and temperature found in actual conditions." This action was too late to affect the certification of the ATR-42 with its early icing-induced control problems. Several months later, ATI had the fatal crash near Como, Italy.

By June 18, 1987, Donald Engen responded to the NTSB regarding the issue that the FAA had previously stated was closed, opened, and closed again and asked for NTSB approval. The letter was intended to get approval for its suggestions to industry. When Jim Burnett responded several months later, in October 1987, his letter went to yet another FAA administrator, T. Allan McArtor. The frustration of the NTSB was evident in the correspondence, which included the sentence, "As different aircraft differ significantly in their response to any one set of [icing] conditions, such an advisory may have little meaning." The advisory referred to FAA suggestions contained in advisory circulars. Also included in the writing were notices to the FAA that the National Plan to Improve Icing Forecasts had been implemented to collect and quantitatively measure atmospheric icing parameters as part of the National Aircraft Icing Technology Plan published in April 1986. This was 13 days before the Como, Italy, crash killed 37.

The banter continued for more than eight years as it had during the previous five years. On September 16, 1994, just six weeks before American Eagle Flight 4184 took off for the last time, David Hinson, the sixth FAA administrator in 13 years to handle this icing issue, moved to close the issue. His letter concluded, and data later brought this into brisk contention, that the FAA's position was that the review of the collected data indicates Part 25 certification rules are "adequate." Hinson proclaimed:

> *The FAA has put in place major programs in recent years which have addressed various anti-ice and de-icing issues. At the same time the FAA has sponsored or collaborated on numerous icing research programs and will continue to conduct programs in this area. However, none of this work has established the foundation or justification to revise [the regulations] as requested in these safety recommendations. While Safety Recommendations A-81-116 and -118 were appropriate when the Board completed its safety report entitled "Aircraft Icing Avoidance and Protection" in 1981, actions taken by the FAA, other government agencies, and industry have addressed these issues over the years. I request that the Board reevaluate the basis of these safety recommendations and classify each as "closed reconsidered."*

The FAA dragged its feet on other safety issues in 1995. Wake turbulence, cyclones of air that trail from the wingtips of aircraft in flight, have caused a number of accidents. The heavier the aircraft, the more violent the wake. This knowledge prompted the FAA, years ago, after the introduction of heavy airplanes such as the DC-10, L-1011, and Boeing 747, to increase separation allowances between a heavy aircraft and a trailing airplane not considered heavy. The designation as heavy is a function of the airplane's maximum takeoff weight.

The Boeing 757 does not qualify for this designation, although the 767 does. The wake turbulence of a Boeing 757, however, is a significant hazard, because it is more violent than some heavier airplanes, having caused a loss of control and subsequent crash of a number of aircraft that follow with standard flight separations. Although the FAA moved to mandate controllers to notify following aircraft of the existence of a Boeing 757 at least 3 miles ahead, this separation rarely provides enough comfort level to pilots familiar with the reputation of the Boeing jet. Pilots might decide to fly high on the glideslope, a dangerous and unauthorized activity, especially with low ceilings, but pilots perceive that as better than wrestling with the unseen twisting winds.

The NTSB reiterated its calls for the FAA to reclassify the Boeing 757 as heavy because of its unusually violent wake and to allow greater separations, although this move would result in slightly less efficient airport operations because the controllers could not pack the aircraft into the flight pattern as tightly. Canada and Great Britain have already acted, but the slow response of the FAA to the issue has caused aviation watchdog groups to enter the fray. To complicate matters, the airline lobby is urging the reduction of separation distances of all airplanes so the carriers can turn the aircraft around faster and make more flights and more profits. The FAA has avoided the issue as long as it possibly can without a huge outcry from an increasingly informed public; it recently established a special office to devise a system to catalog and analyze the data. This move might delay the issue indefinitely.

Another example of FAA management style over substance is the FAA announcement late in 1994 that it would be upgrading the rules covering smaller commuter airline operations to match those of the larger carriers. After two American Eagle crashes of what many people believed to be commuter airplanes because they were propeller aircraft, even though American Eagle operates all of their airplanes to the higher standards, the public demanded action and believed it would be done rapidly because of the FAA press releases. Washington's "during the next 100 days" phrase was invoked for the review of current rules and development of recommended changes. Almost one year later, the agency was barely beyond the planning process. Still to come was a comment period, revisions to the plan, and finally a proposed rule. The process takes years under the FAA's rules, and in the interim, procedures and regulations that can upgrade the safety of the airlines continue to languish or, at best, minor changes are made. With 1.5 million people flying every day, 10 percent of whom are carried on the growing regional airline system, is the traveling public being well served by the FAA?

How can we rely on an agency that has diametrically opposed objectives? The FAA is charged with both promoting and policing the aviation industry. Airlines and aircraft manufacturers are powerful lobbies, and FAA officials often bend to the pressures brought by these organizations. In sworn testimony, an FAA witness stated that, during previous restrictions of the ATR aircraft, "The manufacturer was constantly on the phone badgering them to lift the restrictions."

No amount of public relations work and no number of phone calls from government officials can change the laws of physics and meteorology. The makers of this aircraft have one motivating perspective—profit brought on by the continued confidence and sales of their products. In my opinion, in the face of a mountain of evidence that should have moved them to the contrary, the FAA favored the promotion of the aviation industry at the expense of the safety of the flying public. It was later stated by one aviation authority, who wished to remain anonymous, that the FAA "should have grounded the ATRs long ago, but the U.S. government didn't want to offend friendly foreign countries like France."

During his press conference announcing the lifting of the ban of ATR flight into known or reported icing conditions, David Hinson said one reason behind the decision was the availability of new weather radar technology that allowed for the forecast of dangerous icing conditions, which would theoretically enable airline dispatchers and pilots to avoid these areas. However, pilots being dispatched today are still operating with essentially the same weather data they had before the crash of Flight 4184—the only difference is the addition of icing reports in flight release weather packages. The data produced by the touted Doppler radar for use in interpreting the severity of icing conditions is tough for experienced weather experts to review in hindsight; real-time evaluations by an overworked air traffic control system are nearly impossible. Additionally, it was stated, this technology is designed to detect and identify, in real time, the existence of a weather phenomenon, super-cooled drizzle drops, that was "totally unknown" to all of our experts just weeks before, as indicated by ATR and meteorological experts testifying at the NTSB hearings.

Pilots know and understand that this equipment is years away from fruition and that even when it does come online, it cannot adequately protect the flying public from the variety and violence of weather that nature can produce. In the interim, despite the dramatic announcements by the FAA administrator, the flying public is still in jeopardy.

David Hinson's background calls his credibility into question in the minds of many people. In 1985, Hinson was president and chief executive of Hinson-Mennella, Inc., a Portland-based holding company that owned Flightcraft, Inc., and the Stanley Garage Door Company. That year, he gained control of Midway Airlines, a small but unprofitable regional carrier that operated out of Midway Airport in Chicago.

Hinson purchased another hub for the company in Philadelphia, which began a money drain that could not be stopped before bankruptcy. During the decline, however, the appearance of ethical improprieties tarnished the image of the future FAA chief in media reports. In addition to running Midway, Mr. Hinson was appointed to sit on the board of directors of Continental Bank Corporation in February 1990. By December 1990, Midway was in trouble with the Philadelphia hub, and Hinson was denying representation on the Midway board of directors to Ampco-Pittsburgh, which held 12.4 percent of the airline's stock.

Three months later, Midway announced that it was suspending all lease and loan payments because of financial problems, and Continental Bank disclosed it was a major creditor of the strapped airline. Regardless, the bank nominated Hinson for reelection to the board for the coming year. The Chicago press began looking for impropriety, and soon the airline was in bankruptcy court. One of the first items approved by the court was a $40-million line of credit for Midway from Continental Bank. Hinson stated his position at Continental Bank played no role in the loan.

In April 1991, the shareholders of Continental Bank raised doubts about the continued service of Hinson on the board. The feelings of some shareholders about Mr. Hinson were summed up as "His company didn't go into Chapter 11 by some act of God; his company went into Chapter 11 under his direction." David Hinson said he checked banking regulations to make sure his situation presented no conflict of interest and found none. Under continued pressure, but citing his need to spend more time attending to duties at Midway, Hinson resigned from the board of Continental Bank in July 1991.

Hinson tried to sell the struggling airline, but Northwest Airlines pulled out of a deal that would have saved the jobs of 4300 Midway employees when it found that the management under David Hinson had misrepresented critical load factor and yield information on reports filed with the Department of Transportation. Midway responded by suing Northwest, saying the airline had access to all internal Midway documents in addition to the Department of Transportation reports. In the end, the airline failed to close any type of buyout deal.

By early 1993, the well-connected ex-airline executive landed his appointment to the top position at the FAA. Hinson had gone from flying the skies to ruling them, and it sickened ex-Midway employees who were, through no fault of their own, back at the bottom of the career ladder. Two letters to the editor of the *Chicago Tribune* summed up their feelings at the end of May and early June of 1993.

Thomas Burgan, an ex-Midway pilot, wrote:

There is something fundamentally wrong when people can aspire to positions of power based solely on the connections they have and experience they do not have. I am not a politician. Rather, I was a hard-working professional pilot for Midway Airlines who lost my job 18 months ago.

I am repulsed at the very thought that our ex-CEO at Midway Airlines, David Hinson, could be nominated to such a position of power as the head of the Federal Aviation Administration. It was under his so-called leadership that Midway was put in the precarious position of having to reorganize under Chapter 11. In the end, we filed under Chapter 7, not because Northwest Airlines failed to buy us out, but because it got smart and decided to stay away from a very sick dog.

The safety record that flight crews achieved and were proud of at Midway Airlines was attributable to the intense dedication and experience that each man or woman held to make that little airline work. Hinson was not responsible for my actions, nor those of my colleagues while in the cockpit. Our dedication and professionalism made it work.

These sentiments were echoed by Jim Quinn, who pointed to the perception side of Hinson and press releases when he wrote:

Interesting that another appointment by President Clinton should reflect the composite mediocrity in our government. I'm referring to the person named recently to head the Federal Aviation Administration.

David R. Hinson will be remembered by investors, creditors, bankers, and certainly former Midway Airlines employees for his skillful leadership of Midway Airlines right out of business. And the incident has been treated as an "act of God" rather than pathetic mismanagement. Can the truth really be forgotten so quickly? Maybe your reporters need to look beyond press release-type news to better inform your devoted subscribers.

Mr. Hinson now boasts the active support of Illinois Democrats like Senator Paul Simon, Representatives Dan Rostenkowski and William Lipinski, and even Mayor Richard M. Daley. I would like to know how many of those losers with former vested interests in Midway could be counted on to agree.

Elected officials have proved as ineffective as appointed ones. Many families contacted their elected representatives only to have staff assurances of prompt response and a form letter adapted to respond and close the file. A congressional staffer attending the organizational meeting of the Families of 4184—Roselawn group was shocked when told about the problems still existing with the ATR in ice. "I saw a press conference and I thought that was all fixed," was her astonished response to discussions of the continued problems with the airplane. "The congressman will be interested in talking with you about these issues" was her final remark. To date, we have heard nothing.

If the public wants to protect itself from FAA inaction, or its complicity with big business, then people must put pressure on the pocketbook of the supplier—the airline. No carrier is going to continue to use a craft or operating practice that it does not think is safe, but until the pressure is applied to the bottom line by an alert purchaser, nothing will change. Airlines realize that fact. Comair, a Delta Connection carrier, placed half-page advertisements in *USA Today* and other publications after the American Eagle accidents that built on one simple fact, "Comair operates no ATR aircraft."

14

Families of
4184—Roselawn

The hopes, dreams, and goals of the victims of American Eagle Flight 4184 did not perish when the aircraft fell from the sky on that cold, dark afternoon; because of the enduring love of their families and friends, their dreams are still alive. Something moves people to memorialize the life of individuals taken so suddenly.

A former mentor and colleague of Dana Thompson called her his "shining star" when he wrote to her father about the dynamic young woman's meteoric rise in the cable television industry and the impact she had on him. The Dana Thompson Rising Star Scholarship Fund was founded to provide financial assistance and opportunity for the industry's next rising stars to succeed. The far-reaching effects of Dana's heartwarming smile and hard-driving spirit will help others reach their goals.

Al and Cindy Gagliano established the Jeffrey Gagliano Aviation Scholarship Fund to help new pilots pursue careers in aviation. Jeff Gagliano was doing something he loved when he died. He used to say to his father, "When I'm up in the air, it's like I'm closer to heaven; I can't explain the feeling."

During the summer of 1995, the Gaglianos began planning their annual charity fund-raising picnic, with the proceeds going to the scholarship fund. Their friends and business associates didn't even need to be asked—as the word spread, Al and Cindy were over-whelmed by offers of support. Chefs from all over the Milwaukee area pitched in, portable barbecues were brought to the ranch, and donated food lined two 40-foot tables. Games and entertainment for the kids were set up and all day long, long lines of horses hit the scenic trails around the ranch. A country western band played in the afternoon, and dust from the dirt floor of the auction barn rose into the air, surrounding lines of dancers.

In the chapel, a video presentation played throughout the afternoon. Assembled from photographs and set to music, the short, full life of Jeffrey Gagliano filled the screen. Even the creaking of the floorboards stopped as an almost unending stream of picnic attendees sat as still as statues for the 17-minute show. Songs spilled through the open doors of the small white building into the beautiful mid-September day. The day was perfect, without a cloud in the sky.

The crowd swelled to nearly 700 people, who came early and stayed to the end. Items for the auction were donated by the Green Bay Packers, the Milwaukee Brewers, the Milwaukee Bucks, a local television station, and businesses throughout southeast Wisconsin. Tony Zarinnia and five other pilot friends of Jeff's performed a flyby in single-engine airplanes in the late afternoon and even did a version of the missing man formation in which one plane turns and flies away from the rest of the formation in honor of the lost airman.

Just after the flyby, the goal of $10,000 was reached, and the fund collected nearly $15,000 by the end of the evening. An exhausted Al and Cindy Gagliano thanked guests on an individual basis as they left, and because of Jeff Gagliano, others would be able to realize their dreams of flight.

Delco Electronics set up scholarships in honor of its four employees lost in the accident, held two separate memorial services, and built a small area with benches and brass plaques on pedestals on its corporate grounds in Kokomo, Indiana. The area, designed for reflection as well as memorial purposes, shows that people in large corporations can also be caring individuals.

Dick Cunningham spent his entire adult life at Delco, starting as a mail clerk in 1962, when he was a 19-year old graduate of Galveston High School in Indiana. By 1994, at the age of 51, Dick was the manager of new ventures for Delco Electronics and a leader in the effort to expand his company's worldwide operations. Family was important to Dick, and his family found out how important Dick had been to hundreds of his coworkers, as people stood in line for more than two hours at his funeral to express how they felt about him to Beverly, his widow.

John and Toni MacDonald both worked at Delco but had first met on a blind date set up by coworkers; on August 11, 1990, they were married. Three years later, on August 18, 1993, John Joseph III was born. "In every way, we were living our dream," Toni MacDonald remembered in a letter after the accident. When John kissed his wife and son goodbye before he boarded the flight, 14-month-old J.J. pointed to the cloudy sky and airplane on the ramp and urged, "Da Da, plane, no, no."

David Snyder related a story of life to his son when an illness threatened the life of Steve Snyder's grandmother. "This is the way things have always worked," David said. "People are born, they live their lives for a while and then they die. It has always worked that way and will continue; you just have to go on." That pragmatic view of life enabled David to break down any problem and analyze and solve it. The instructions for assembling the toys on Christmas never proved a challenge for the ingenious electrical engineer who headed the international operations support program for Delco. "He could even make the toys work with parts missing," his sisters said.

Brad Stansberry was the antithesis of Will Rogers; he never met anyone who didn't like him. Although only 27 when he died, everyone remembered him as making a difference in the lives of other people by what he gave them. His willingness to take on any task also made him an asset to Delco. More than 1000 friends, including members of the Madison Heights marching band and the Star of Indiana Drum and Bugle Corps, with whom Brad had once played, gathered in a gymnasium to remember him. On the shiny wood floor, the hats he earned in both associations were placed next to an empty music stand beside one of the many musical instruments he played. Brad cared deeply about his family and friends; they repaid the debt to his mother, father, sister, and brother by their caring in return.

Barbara Tribble had volunteered for the Austin Street Shelter in Dallas for a number of years. She never sought recognition for her actions and was remembered as helping in the shadows. Even her family did not fully realize how Barbara had committed herself to the plight of the homeless and others helped at Austin Street. Mobil Oil, Barbara's employer, set up a memorial fund in her name so the work that meant so much to her would continue. One of the recipients of aid from the Austin Street Shelter summed it up by saying, "There is caring here; not just from the people that help but coming from the other side as well. We do care, too."

Jose W. Calderon began helping his community at an early age and hoped to continue the work after beginning his job with a Chicago law firm. Jose tended a garden at the 4-H Club in Lincoln Park in Chicago when he was only 12. Just blocks away from skyscrapers and busy streets, he and other youngsters learned country skills. Most people believe the 4-H Club is for country children, but the club that began more than 90 years ago to provide rural children an opportunity to use their heads, hands, hearts, and health found a home in the zoo at the Chicago park. The setting of the club was a long way from an office in one of those skyscrapers, but Jose was on his way. He had been sworn in as an Indiana attorney and 11 days later would have taken the oath in Illinois. Eva, his mother, attended the Illinois ceremony along with 1500 others and accepted a special award from Chief Justice Michael Bilandic of the Illinois Supreme Court. Her husband, Jose V. Calderon, also died on the flight.

Lee and Bernice Stackhouse gave back some of their business success by giving of themselves. The couple had raised their children and had new grandchildren arriving but found time to take a 14-year-old distant relative under their wings. Javier Sayles' father worked for Lee Stackhouse in Los Mochis, Mexico, and Javier was traveling with Lee and Bernice when the flight went down.

Friends remembered David Shellberg by having a service and serving his favorite chocolate chip cookies at the reception afterward. David had kept people together as passing years worked to separate them. He sent letters and cards to update friends and former teachers of what was happening in his life, but he also was intensely interested in what was going on in their lives.

Frank Sheridan had a zest for life that encouraged him to push the human envelope. Most people would be content with one of Frank's adventures or achievements, but his life overflowed with them. He exhibited the drive and responsibility endemic of being the eldest son with five siblings, and to each one he was a confidant, friend, and adviser. The diversity of his life experiences brought a perspective to any conversation that made his role as a mentor irreplaceable.

Frank became an Eagle Scout and exhibited his heroism in Vietnam as a medic in the U.S. Marine Corps. He went out under fire to recover wounded or dying soldiers and earned two Purple Hearts and a Bronze Star. His exposure to death gave him a thirst for life and a deep appreciation for family and friends. Frank took an intense interest in the lives of his younger brothers and sisters. When nieces and nephews came along, they gave him the personal satisfaction he never achieved by having children of his own. He attended the youngsters' sporting or social events and took pride in their achievements as they did his. Frank was as quick with a compliment as he was with a well-intentioned suggestion.

Friends and strangers who met Frank knew they had been with someone special because he made people believe they were the most important person to him at that time. One of his siblings would write that, "Everyone was a thread in the fabric of his life."

Frank explored the entire world during his 48 years. He kayaked on the Pacific Ocean, with his photograph winding up in *National Geographic* magazine, flew around the world in an open cockpit Stearman biwing aircraft, and searched for World War II airplanes in the jungles and waters surrounding New Guinea. An excursion to

New Guinea yielded underwater and land discoveries of downed or abandoned Japanese and American aircraft that were vividly described in an article authored by Frank.

Flying an open cockpit airplane around the world had not been done since 1924, but Frank joined a team of adventurers from southern California who formed a group called The Northern Light and set out on a 23,000-mile trip, leaving and returning from San Diego. The name came from the route chosen for the journey, which took them to Russia via Canada, Greenland, and Iceland, returning over Scotland, England, France, Germany, and Poland.

Brad Stansberry's younger sister, Jennifer, was a 21-year-old college student at Purdue University on Halloween 1994. In the aftermath of the accident, she began to see through the misdirections being perpetrated by the airline and spoke out publicly, rapidly becoming a regional media personality. She voiced details surrounding issues that American Airlines would have rather kept under covers. She informed the press about the constant problems her family was having in obtaining accurate facts about the accident, passed on the existence of the mass grave and what originally was called a "midnight burial," and led a march against the inability of families to obtain any of the personal effects.

Doug Smith worked quietly behind the scenes. The father of Alison Field knew the real power for change was located behind the closed doors in Washington, D.C., if he could get past the form letter responses of politicians. Working with the elected officials and appointees that headed the agencies would not be easy. Many of the families had written to politicians, and the response often was that a congressman or senator was concerned about the issues raised in your letter and he or she would be investigating the matter.

During the summer of 1995, Doug became president of a new national association of families involved in aviation disasters, the National Air Disaster Alliance. The group met with heads of agencies who in turn met with airlines, and the group contacted congressional staffers in hopes of reaching representatives or senators. I discussed progress with Doug recently, and he indicated it would take three to five years to make any substantive changes; however, he was not going anywhere before the job was complete. He told me we cannot alienate everyone in the agencies because if Bill Clinton is not re-elected in 1996, David Hinson of the FAA will be gone, Secretary of Transportation Federico Peña will be out, and Jim Hall of the NTSB may be replaced. The entire process would start again with new faces. Doug works for change in the future.

After the hearings in Indianapolis, where many of the families met for the first time, people began to interact with one another and realized the need for an organized effort to accomplish several simple but meaningful goals. Terri and Tom Severin became comfortable discussing the issues with the media in a conservative, thoughtful manner and verbalized the ideas of Doug Smith. These three were prepared to lead the Families of 4184—Roselawn organization when it was formed.

Not every family interacted at the hearings; some simply wanted to put the ugly event behind them. They watched from a distance or quickly exited during breaks in the proceedings and appeared to keep a distance from the vocal dissertations of the others.

A few telephone calls had been made to the Aguiar and Gagliano families. The calls were ugly and served no purpose but to cause more pain. The callers, who did not identify themselves, referred to the pilots as the killers of their loved ones. After a newspaper article within weeks of the accident in the *Washington Post*, most likely the result of early ATR rumors, claimed that the crash was the result of pilot error, the telephone calls increased. Relatives of the pilots did not know the details of the accident; the company and even ALPA kept them in the dark as much as they did the other families. Everyone approached the hearings and other families cautiously. Al and Cindy Gagliano were a jumble of nerves before the hearings began because of the uncertainty of the outcome and the potential for confrontations with other families.

Fortunately, no confrontations occurred with the families of the pilots, although there were furtive glances from some relatives of other victims. The cautious introductions during the first few days flowed into a friendly interest and later the trusting exchange of emotions that sealed the bond. Small groups of family members were seen talking quietly in the lobby of the large meeting room, showing treasured photographs to one another and displaying an honest interest in each other's loss. For the most part, by the end of the week of hearings, communication between families was good as they swapped phone numbers and promised to stay in touch.

Tom and Terri Severin led the charge to get the group together and worked feverishly to plan an organizational meeting of the families to identify concerns, plot a preliminary course of action, and strengthen the network of relationships that began in the last days of February 1995.

The first meeting was on April 30, 1995. The group met for the organizational meeting in Glenview, Illinois, on a drizzly morning that turned into a clear but cold afternoon. The meeting began informally and sputtered early but took on a dynamic of its own by the afternoon. By the time the 5:00 P.M. deadline came for the closing of the donated motel meeting room, it was apparent nobody wanted it to end.

The group selected officers and a name, "Families of 4184—Roselawn" and identified major goals of the group. Linde and I watched from a comfortable distance and were joined in the afternoon by Tony Zarinnia. We were the only nonfamily members invited to the meeting. We participated in the discussions but felt it best for the family members to direct grievances from their perspectives. We were there as support, not active participants, and I recalled my friends as I listened to the discussion.

I think of these families every day. Not a day has passed since the crash of Flight 4184 that I do not endure periods of melancholy over this tragedy and bittersweet remembrances of the people lost to us all. But I am always reminded that I am at the periphery of the human destruction. While I shall forever treasure the memories of my three friends, how close were we really?

I flew with Captain Orlando Aguiar on a number of occasions, and we shared hours of idle conversation. Orlando and I got along well in the cockpit, and I found his easy way of running the flight deck and aircraft a comfortable way of operating and hoped to use what he taught me when I had the chance to move to the captain's seat. More can be accomplished between professionals in a relaxed manner than under the stress of overbearing dictatorship.

Flight attendant Sandi Modaff worked my first airline trips as our third crewmember. She was friendly, smiling, and bubbly. I still smile at how she had irritated me one day by indignantly asking, "Who made that lousy landing?" My reply was an embarrassed "I did." I did not explain that the reason for the rough arrival was a wake vortex from the previous aircraft on the runway. She would not have cared anyway. She had already turned on her heel and left the cockpit, leaving the captain and myself to shake our heads while he snickered and I fumed. She made up for injuring my professional pride by treating my parents like royalty on a short flight from Milwaukee to Chicago O'Hare. I mentioned the fact that I had arranged to fly the flight because they would be on board, and she took it upon herself to identify them and make sure they were extra comfortable.

First Officer Jeffrey Gagliano and I were hired on the same day and went through our training together. We spent hours studying ATR aircraft systems and operations materials while in ground school in Marquette. Five of us selected for ATR training out of our class of 30 pilots became an enclave within the group, secretly relishing that we had been chosen for the larger, more sophisticated, airplane over our contemporaries.

I shared meals with Orlando, Jeff, and Sandi on occasion when our schedules allowed and always shared nodded acknowledgements and smiles as we passed in the terminal. I must remember that I was only a friend from work—an acquaintance of circumstance. If I have these deep emotions about this experience, how must those who were family suffer through each day?

That question has been answered many times during my meetings with family members at the hearings in Indianapolis and afterward. Each asked the same questions about the accident, the pilots, and the airplane; even with answers, many still seemed to have the same emotional emptiness in their lives. Perhaps the group support of the Families of 4184—Roselawn might allow a greater chance of closure for these people. Some were not yet ready for participation in the group, and others were separated by distance or culture but yearned for information.

Although the family discussions at times digressed to individual stories of lost loved ones, Tom and Terri maintained the focus needed to cover the informal agenda in the allotted time, gently prodding the meeting along when it got bogged down. A clear set of issues emerged as common ground by the afternoon. Clear agreement was reached with regard to the ineffectiveness of the CARE team members in providing proper and timely information to the families, as well as a deep concern over the confidentiality of the discussions held between families and airline employees.

The return of personal effects was a major issue with the families in attendance. Relatives who did not realize the level of destruction in the accident were delicately informed by others in the room. Stories were told about caring recovery workers who had contacted people from the group to inform them of large amounts of personal effects that were recovered and cleaned, but every person in attendance at the meeting in April had received no more than three items from the airline for each of their lost relatives.

Three seemed to be the magic number. Sometimes it was a credit card, driver's license, and a photograph cut from a passport; other times it was an identification card, set of keys, and a laminated photo; but personal effects always seemed to be returned in groups of three. Lists of effects identified to belong to certain victims had been made, but the airline had destroyed whatever it felt was not important under the guise of it being a biohazard. The attendees at the meeting reported conflicting stories that originated from American Eagle, which added to the confusion and anger.

Tony Zukauskas was at the meeting representing Eileen Novichuk. Tony told the group he worked in the dispatch department of TWA and that his airline had a shipment of checks on American Eagle Flight 4184. While the families waited for months only to have a small number of effects returned, Tony reported that American Eagle had returned the shipment of checks to TWA within days of the accident. No mention was made, according to Tony, of any concerns about the biohazard. Relatives wondered why paper checks were returned while the intact and identified day planner of Frank Sheridan was destroyed without notifying his family.

The cleanup company stated that American Eagle had wanted the recovery of the "high-ticket items," and the rest could be destroyed. Relatives at the meeting discussed the story that was on *48 Hours* and were disgusted by the playing to the cameras American had done in returning the intact belongings of a crash survivor while failing to respond to the requests of a family member of someone who had died. Just like Katie Hales, the families of Eagle Flight 4184 received unannounced packages brought by an unsuspecting and smiling package-delivery employee. Someone noticed the unusual return of a roll of Lifesavers candy to Katie as not being in line with worries of a biohazard. Apparently the deceit was not universally applied.

After the discovery of the mass grave and the news that American Eagle was relenting to pressure and had agreed to place a "sterile" monument on site that would not list victim names, hometown, age, or any mention of an aircraft accident as the reason for the mass grave site, the discussion was lively and agreement was unanimous with regard to continued pressure on American Eagle to provide personal information on the monument.

We sadly acknowledged that the field had already most likely been sown by Clarence Hanley, and we had lost that fight. Although the pain of knowing what was still in that field caused some tears, some relatives who were pragmatic spoke of returning to the earth from which we are made.

The night before the meeting, a storm passed over the Dallas-Fort Worth Airport, the headquarters of AMR Corporation and main hub of American Airlines. Beginning as a weak cold front, the storm built rapidly with warm, moist air being forced aloft, creating hail 3 to 6 inches in diameter. Nearly 80 aircraft were damaged when the hailstorm pummeled the airport. Fifteen airplanes belonged to Delta Airlines, which quickly returned them to service, but the rest, representing nearly nine percent of the American and American Eagle fleets, were severely damaged and out of service for a month or more. Because of the damaged airplanes, American Airlines canceled 219 flights on the day we held our meeting; no one commented on the news, but several people smiled.

When the meeting adjourned, about a dozen of us went to eat at a local restaurant and discuss the future of the group. Afterward, on returning to the hotel, I brought up the subject of this book and my interest in telling the story of the tragedy from both a human and technical aspect. The idea was greeted with enthusiastic promises of support and the hope that a fair and accurate telling of the story would help generate the changes needed.

We had a battle on our hands against a well-funded, well-connected group of companies that could hunker down and weather the media attention, which they knew would wane. It could be a waiting game on their side while we fought for attention to issues that were important and urgent to us but old news to an increasingly indifferent public. The focus had to be on the future with a reference to the past. American Eagle could not forever be called to atone for what it did to us; however, the desire was to help those who inevitably would come later.

No help would come from American Airlines or any of the AMR companies; even the crew was not honored with any type of memorial fund set up by the airline. Members of the Air Line Pilots Association (ALPA) would spend more time slapping each other on the back and offering congratulations to each other for their work on the icing problems and family issues, but the families of both pilots did not even hear from the union for months at a time.

Anything accomplished to honor the memories of the loved ones was a product of the work of family members who often were too grief-stricken or financially strapped to help as much as they wanted. Telephone bills soared and travel expenses mounted, but the chain that was forged during the days and months after the crash bound the diverse group together and moved them forward.

The feelings were summed up in an anonymous response to a polling of family members that took place in May 1995. Although there were opposing viewpoints, the following letter addressed only "To Whom It May Concern" seemed to capture the attitude of most people in the group.

My husband was killed in the American Eagle crash of Flight 4184 in Roselawn, Indiana, on October 31, 1994. Naturally, it has been a shocking and traumatic experience for me.

However, I resent the insensitive treatment I have received from the airlines.

When the crash occurred, they made me wait 24 grueling hours for official confirmation that my husband was on that flight. I think that was very unnecessary torture. They made it virtually impossible for us to get any information thereafter other than what they chose to disclose to us through a very inexperienced "Care Person."

This Care Person asked us to describe what jewelry my husband was wearing to help in the identification process. They took two weeks to identify him.

As far as his jewelry and other personal effects, I have yet to receive a claim form from the airlines. I have been told all personal effects were destroyed by the airlines except what they considered valuable. How dare they!

When I inquired in November 1994 about when I would receive my husband's personal effects, my Care Person said not until after January 1, 1995. Shortly after January 1st, I called my Care Person again inquiring. She said no personal effects survived the crash other than metal objects such as jewelry and they had to go through a sterilization process. She also said no belongings of my husband's were recovered. Then I received a letter from AMR with a list of objects, mostly jewelry, and it included a description of a watch like my husband's. I gave my lawyer the receipt for the watch but I have received no communication at all. They also sent a list of every credit card my husband had in his wallet including personal cards. Where are they and where is his wallet?

On March 27, 1995, I requested a Coroner's report on my husband from the Newton County Coroner, Dr. David Dennis. I was told there was none but he'd write something up. On April 10, 1995, I got a call from Dr. Dennis' secretary saying he is working on the report and will mail it shortly. I have not received anything to this date. It took American Airlines two weeks to identify my husband. We gave them dental records and fingerprints, as they had asked. They sent home a very small amount of remains. In February 1995, we found out that Dr. David Dennis, the Newton County Coroner, did not bother to take the time to identify a very large amount of remains. He instead authorized them to be buried in 19 caskets at the Calumet Cemetery. I am hurt and very angry that the airlines allowed this to be done in secrecy with total disregard for the surviving families' grief and right to know the truth.

After much embarrassment and pressure, AMR has agreed to erect a monument at the grave site at Calumet Park Cemetery, but they have refused the families' request for ages and hometowns to be inscribed on the monument alongside each victim's name. I don't think this is an unreasonable request after what we have been through. Please address the insensitive treatment we, the Families of 4184, have received so that other families do not have to endure the same in the future.

Anonymous
May 29, 1995

15

Drawing the legal battle lines

Negligence; willful negligence; intentional negligence; wanton negligence. The terms will possibly only be used on papers filed in courts across the country. Realistically, the words will never be spoken in front of a jury in connection with the case of the victims of American Eagle Flight 4184. In most instances, according to a study completed in 1988, the families of victims or survivors of airline crashes settle their damage suits against the defendants long before the parties get close to a courtroom. They settle for pennies on the dollar.

Families do not function cognitively for a long period after the loss of a loved one and are therefore vulnerable to inadvertently providing ammunition to the airline through its CARE team representatives. Allegedly, after a number of previous accidents, airlines acquired knowledge that relatives insist could only have come from discussions with the volunteers from the airlines. Individuals with the pain of recent loss naturally reach out to a caring individual during times of stress, and the stages of grief are well understood by the trained airline representatives. Anger and desire for justice come long after the initial stages of shock and disbelief, and long after the CARE teams have been recalled by the airline. American Eagle and most other airlines deny any connection between the volunteers and the legal or insurance people, but consider if you would want employees of the company whose aircraft just had a disaster to be in your home during a time of possibly muddled thinking. People have a right to control the situation and refuse the entry of the CARE team into their homes and lives, but few have the strength or foresight to make that choice in such a circumstance.

When an airliner crashes, airline executives realize they have a serious problem on their hands. But in careful thought, they realize that insurance coverage will protect the airline from much of the financial impact, and, in the case of American Eagle, a $1-billion umbrella protects the company. Plaintiffs who eventually enter the anger stage of grief believe they are hurting the airline with their claims, but in reality the financial loss to the airline, aside from loss of revenues—which the carrier also might have covered by the aircraft manufacturer in later deals—will be minimal.

Insurance adjusters know the situation. The airline has paid them hundreds of millions of dollars during the past decade, and much of the payout would be against the insurance. However, with no survivors, there was little chance of high compensation, even if the cases got in front of juries. For the airline, it was better financially that nobody survived. Just like auto insurance, after an aviation accident occurs, the underwriters insist on higher premiums.

The public perception is that everyone losing a relative in the accident will be immediate millionaires. Media coverage of awards for damages in high-profile cases of a million dollars or more make headlines and stick in the minds of the public, but those are not the norm. In truth, if the award is large, the case will be appealed. Most families will deal with the pain for years before a dime is paid to them by the defendants. A family that lost the sole breadwinner or a spouse who lost his or her mate must now seek special care for his or her youngsters and might struggle financially for most of a decade. The possibility that relatives might also become victims is as likely as the chance of obtaining justice.

To determine the historic level of compensation to victims of air disasters, a study was commissioned by a group including trade associations for airlines and aircraft manufacturers. The report was released in late 1988. RAND's Institute for Civil Justice did the research, organized the statistics, developed opinions, and offered its conclusions to the public and sponsors, some of whom disagreed with the findings. According to the report, "Relatives of people killed in airline crashes are paid less than half their economic losses, and the legal battles give little incentive for improving safety." After paying more than $400,000 for their share of the $1.3 million cost, trade associations for the aircraft manufacturers and airlines protested the conclusion, which they said should not lead the public to conclude "carriers are flying aircraft they know are unsafe because they know their [financial] exposure isn't great enough to warrant additional care."

Economist James Smith pointed to the inadequacy and inequity evident in the current system when he stated, "It's not adequate in that people don't receive full compensation for their loss. It's not fair because people with identical losses receive very different amounts." The group studied the 25 worst air disasters, which claimed 2228 lives between 1970 and 1984 and discussed many disturbing situations that are still in effect today.

For example, travelers flying on an international ticket, even though the accident may have occurred during a domestic flight, are limited to a liability cap of $75,000 under the terms of the 1929 Warsaw Convention. Thus, any passenger flying on Eagle Flight 4184 who was connecting to a flight out of the country might have their families compensated at a lesser amount compared with those whose tickets were for travel entirely within the United States.

This liability limit of the International Civil Aeronautics Organization treaty has been challenged, and if plaintiffs in this case can prove the airline guilty of willful, wanton, or intentional negligence, they have federal precedent for seeking additional damages. The court ruling came after Korean Airlines' Flight 007, a Boeing 747, was shot down by a Soviet fighter plane on September 1, 1983, and 269 people died. A jury awarded $50 million in punitive damages because the airline failed to correct a navigational error that placed the flight more than 300 miles off course and in Soviet airspace. A similar ruling was previously adopted by the Second Circuit Court of Appeals on December 21, 1988, when the Pan Am crash over Lockerbie, Scotland, was ruled negligent because of lax security.

The RAND study also found another inequity: Most surviving relatives only received $363,000, or about 49 percent of the $749,000 true economic loss as a projection of the amount that the victim would have earned if he or she had lived. That amount is dependent on the earning capacity of the victim, which resulted in about half of the dollars paid going to one-sixth of the victims' families. For all 2228 deaths during the studied period, $769 million was paid by airlines, aircraft manufacturers, and government agencies, compared with an economic loss of $1.59 billion, or $345,153 per life lost.

Attorneys in civil lawsuits seek compensatory damages and, depending on the nature of the claim, punitive damages. Compensatory damages serve to compensate the victim or relative for losses such as pain and suffering and loss of companionship, services, and society; medical costs; and loss of earnings. Punitive damages are intended to punish the guilty party by allowing the judge or jury to award an amount in excess of compensatory damages. To prevail for punitive damages, a plaintiff must be able to prove compensatory damages. State laws vary widely on the limits of damages, both compensatory and punitive, so the location of determined jurisdiction in an airline accident is almost as important to the parties as the issues at hand.

Most aviation-disaster litigation is settled before trial, and some families are compensated without needing to file suit because an insurance company is eager to get the matter behind it as quickly as possible. After accidents in the past, insurers often offered more money initially, before families file a lawsuit, than they do afterward. If a lawsuit is filed, the settlement offer normally decreases.

The issues in the courtroom are most often limited to those surrounding damages, because the defendants often admit their responsibility, therefore eliminating any debate on the cause or the circumstances of the crash or deaths. When defendants admit responsibility, the plaintiffs' attorney is precluded from developing any testimony beyond the "value" of the victim. The tactic works for defendants, who wish to avoid gruesome and often sensational evidence that might sway jurors to award higher compensation or drag the accident to the forefront of public attention.

Although lawsuits can be filed as soon as the day after the accident, the final resolution of the case sometimes takes a decade. Each time the family responds to another deposition request, meets its attorney to discuss the case, or reads an article about another lawsuit that is settled from the same accident, the nightmare comes back. People who have begun to get on with their lives are dragged back to the pain that may be as fresh as the day of the accident. Airlines and insurance companies know that. This victim abuse can be an effective tactic for moving a particularly difficult case to settlement.

Lawsuits in the case of American Eagle Flight 4184 are likely to be brought in state and federal courts throughout the country in cities where the various victims lived, in Indiana where the aircraft crashed, or in any location where any one of the AMR subsidiaries is quartered. When a foreign manufacturer is included in the suit, some cases are filed in various districts of federal court. The issues to be decided in the matter of the Roselawn accident are numerous, complicated, and very fluid. A jury trial is often more beneficial to the plaintiffs, who hope to tug the emotions of the jury. A bench trial may yield only the cold, unflappable jurisprudence of a trained and experienced judge. The variables of the options are well known to the attorneys of both sides, and the attorneys for ATR immediately petitioned for a removal of suits from the crash of Flight 4184 from Illinois State Courts, which use a jury, to federal court, in which a judge hears the case on the grounds that the client is a foreign entity. In the end, complex and furious legal wrangling will determine the locale of each of the cases that proceeds.

Locale can be as critical to the required legal maneuvering as it is to the amount of awards. No question exists as to whether the defendants should compensate the plaintiffs, but in the legal chess game, planning must be done several moves ahead for an overwhelming win. Even if an attorney expects an out-of-court settlement, he or she must prepare for the trial if an agreement cannot be carved out. Although some attorneys may be eager to try the case to right the injustices perpetrated by the defendants, they are bound by the wishes of their clients.

Because of the large number of damage claims in an airline accident and the similarity of the nature of the claims, cases that are not settled are sometimes combined into a single trial, although a more common occurrence is the combining of the discovery process under the jurisdiction of a single court. The decision on the consolidation of either the discovery portion or the entire process leading to trial is often a controversial matter as attorneys lobby for the best interests of their clients.

In the case of Flight 4184, the courts ruled that the cases would be combined for purposes of discovery under the auspices of a court in Chicago so that the time of both the court and attorneys for the defendants, who would be traveling all over the country with their clients for up to 68 different cases, could be used most effectively. In late October 1995, a hearing was held to determine the scope of the discovery to be allowed in the cases. The insurance carrier for AMR, represented by the same Robert Alpert from the aborted visit in front of the *48 Hours* cameras, indicated the defendants could stipulate to the liability, which would eliminate any in-depth discovery of the technical facts surrounding the accident and preclude the plaintiffs from ferreting out the truth surrounding the accident. The court's decision was delayed until November 16, 1995, when Alpert would be asked for his decision about admitting liability.

As of late April 1996, no admission of liability in the crash of Flight 4184 had been formally proffered before the Court. As parties continue to maneuver for advantage, ATR is still attempting to extricate itself from lawsuits, claiming status as a "foreign sovereign," which would make the manufacturer unanswerable to claims filed in U.S. courts. As of this writing, the United States 7th Circuit Court of Appeals has yet to rule on this crucial issue. The continued delay in the ruling is causing several involved persons to wonder about the potential effect on the Court of the rumors of political pressure from the French. The NTSB reported that even the White House had been pressed by French government representatives for help in the matter of the content and tone of the final accident report being prepared by the NTSB technical staff. This information, released in early April, can only discourage those awaiting the Court's ruling.

According to Captain Aguiar's widow, Dawna, agents of AMR and ATR attempted to separate her from the rest of the cases by threatening to drag in the couple's personal life. During his tenure at American Eagle, Orlando Aguiar had a reputation for being a ladies' man—the type of reputation rarely backed up by fact. Aguiar was divorced from his first wife, and he later met and married Dawna when she worked as a flight attendant for the airline. The birth of his daughter, Marisa, gave Orlando great pleasure. He and Dawna found out that she was pregnant again shortly before the accident, and the couple worked through bumps in the road like any married couple. Dawna said she was in a state of despair when an agent for AMR and ATR reportedly threatened her with public disclosure of private matters that had no bearing on the accident by stating they had "two Simmons Airlines employees prepared to testify to overhearing one side of a telephone argument."

Dawna had heard the gossip at the company and knew how vicious and hurtful, true or not, that it could be. She had lost her husband in a violent and unexpected accident; she had carried and given birth to a baby girl on June 9, 1995; she said she had been continually forced to badger the worker's compensation and health insurance carriers for AMR to process payments for support and care of her two daughters promptly; and she even complied with intrusive requests for disclosures of intimacies concerning her married life with Orlando. Dawna had held up well to the harassment, but when the company to whom her husband had given his life in service indicated it meant to tarnish the only thing left they had not destroyed, the preservation of his reputation in the eyes of their two daughters took priority in Dawna's mind.

In early December 1994, within a week of my name becoming known publicly as the rebel American Eagle pilot, correspondence from attorneys representing families of crash victims started to arrive at my home, asking for my help. I had already committed to provide any technical information necessary to help the relatives obtain justice.

At that point, nobody knew the positions ATR and AMR would take in defending the lawsuits being filed in late November. It later became clear that both companies would rely on a similar philosophy: "The FAA made me do it." AMR stated that the ATR is certified by the FAA and that it is not incumbent on them, nor do they have the resources, to recheck any of the available data on the aircraft. They would not make changes until the FAA told them the aircraft needed changes. It was similar with ATR; the aircraft, it said, met all the requirements of FAA certification. If it meets the FAA requirements, the company has no responsibility to go beyond the minimums. That is why the final NTSB report will be important to both companies.

In the culmination of the accident investigation of the ATR near Como, Italy, the hottest debate took place over the degree of icing that the aircraft encountered before the pilots lost control. If the report concluded, as the Italian weather service had determined, that the icing conditions were moderate, ATR would be in trouble because the aircraft is supposed to be able to operate in moderate icing. If the report said, as ATR demanded, the icing conditions were severe or extremely severe, as reports to the United States from the French indicated, the aircraft's reputation and culpability was out of danger because the pilots would have been guilty of taking the airliner into conditions that exceeded its operating limitations. The question would be the same in the aftermath of Flight 4184.

Remember that aircraft icing conditions specificity or type is a factor. What may be severe icing exposure for a Piper Cub may be trace exposure for a Boeing 747. With the ATR's history of problems, it would also be proper for the pilots to think of the ATR as having a sign stating "do not operate in ice" on the instrument panel. Of course, that would not do for the airline.

On October 23, 1995, Anthony Broderick, associate administrator of the FAA, reported the agency had almost completed testing of other turboprop regional airliners to determine whether they had the same aversion to ice the ATR displayed. According to the *New York Times*, Broderick stated that some of the others share some of the vulnerability of the ATR to icing, but none is as vulnerable. That type of information hurts ATR's position in the legal battles to come and in the competitive marketplace. After all, why would an airline purchase a particular aircraft that has problems when a competitor's product does not and performs the same mission? In Europe and other parts of the world, airlines have learned from the lessons of the history of the ATR aircraft, and some have moved away from the aircraft. One attorney concluded, after researching the ATRs, that "worldwide, the ATRs are being moved to warm weather operating climates because of the icing problems."

The court trial will be determined by the statements and credibility of the experts. Immediately after the accident, both sides attempted to secure the foremost authorities in the world; ATR even sent a check in a letter to Porter Perkins, the icing expert. The purpose of the check was not to secure his help but to keep him from the plaintiffs. If Porter Perkins, described by NTSB hearing expert Marcia Politovich as "one of the fathers of icing science," cashed the ATR check, he would been "buried" by the aircraft manufacturer. ATR would not have used Perkins itself because his views would most likely contradict its claims. Fortunately, Perkins held that check. He stated later, after being hired by an attorney representing nearly 30 percent of the families involved, that he had been waiting for the "right cause to come along in this case." Should cases go to trial, Perkins will be a credible, forth-right witness who is dangerous to ATR.

Offers of payment came for my information and help; however, this was a special situation to me, and I made that fact clear to any-one who called. I was committed to help the families in any way I could without payment. I am doing what I think is right.

This event has been costly beyond monetary terms for thousands of people, including the families and friends of victims and the airline employees. Losses are covered by insurance, but damage has been done to the reputations of the ATR aircraft and American Eagle. In ad-dition, people have come to believe that the system is unsafe.

Regardless of the damage done, an attempt will be made to place a value on the lives of 68 individuals. Families will attempt to put the nightmare behind them, and some will settle their claims early to move on with their lives. Already, one family has reportedly settled on $1.3 million for the loss of a wife and mother of two children. The husband will get a third, the sons will divide a third, and the lawyer will be paid a third. The money deemed to compensate a little boy for the removal of his mother from his life was tabulated as $216,666—less money than the insurance companies historically paid for the loss of life a decade ago. This family's decision could affect each of the other claims in the crash; however, in the end, this family might be the lucky one, closing the issue early and allowing the pain to heal. Others will go on for years and still might wind up with similar compensation. Even if a fam-ily bears up under the pain and sacrifices needed to pursue its case to a conclusion in court for 6 to 10 years, and they prevail with an ac-ceptable financial award, chances are the defendants will file an ap-peal. The decision then needs to be made whether to continue the fight or accept a settlement.

Should the cases drag out past the turn of the century, individu-als involved in the original tragedy will have died, retired from, or left the employ of the companies responsible, and the problems will be shouldered by people not connected to the accident. AMR Eagle president Robert Martens and Simmons Airlines president Peter Piper have already left to pursue other interests, as have other lower-level individuals. AMR chairman Robert Crandall will retire during the next five years with lucrative retirement benefits and hearty congratula-tions for his service to AMR Corporation, whereas the relatives of the victims of Flight 4184 will still struggle for answers and justice.

16

Assigning a cause

People in aviation around the world hold their breath when the U.S. National Transportation Safety Board meets to issue its official findings on an aircraft accident; two simple words, "probable cause," can have tremendous ramifications to those involved in the industry. Relatives of victims killed in the accident often look to the official assignment of a cause as hopefully the final step in a long and painful journey. For the pilots' families, it can be a strenuous time just before the public determination of cause because the pilots' judgments placed the craft into the situation in which an accident occurred. In the case of American Eagle Flight 4184, the families were dragged along as unwilling passengers on an emotional roller coaster while the date for the final hearing was continually postponed.

The date of the announcement of the cause of the crash of Flight 4184 was originally during the summer of 1995, but by July, when I traveled to Washington and visited with the NTSB, the date had already been moved to early October. Greg Feith, the investigator in charge, was writing the report as I sat in the office of NTSB public affairs officer Pat Cariseo. Cariseo told me the date would be moved back to late October. It would still be less than one year to bring the matter to conclusion, I thought. The timing would mean one large emotional jolt to the families would come at the end of October, but then they would be able to attempt to move forward with their lives to some degree.

Before I left Cariseo's office, I asked whether it would be alright to speak with Greg Feith about the investigation after the hearing in October. We had agreed it best that I not meet with him before the hearing to avoid any allegations of bias. When I walked out of the meeting, I was glad for the information I had received, and I made sure the word was passed through the family grapevine.

For American Eagle, the outcome of the investigation of Flight 4184 could result in a severe economic impact on its operations. American Eagle had already been damaged by media reports of the pilot revolt precipitated by the mishandling of previous incidents that many said predicted the accident; improper crash site cleanup; and the alleged insensitive treatment of the families. These factors, when combined with the general aversion to propeller aircraft airline travel and the required shuffling of the fleet to accommodate the FAA flight ban, resulted in a decrease in passenger bookings of more than 7 percent in the first quarter of 1995. American Eagle's total revenues were down 31 percent in January 1995 when compared with the previous January, and first-quarter revenues reportedly were expected to drop by 21 percent from the first quarter of 1994. Complicating the problem of less revenue, the huge cost of relocating the ATRs to southern climates and associated crew transport and lodging added nearly 19 percent to the expense factor of each available seat mile. American Eagle was hurt, and even though supported by the wealth and resources of the parent AMR Corporation, the stockholders might have been upset by a further slide.

Even though the carrier had more than adequate insurance supplied through AMR Corporation, it had to come up with an additional $15 million for its liability insurance for 1995. The annual policy had been renewed just 12 days before the second fatal crash of the year for the airline, on December 13, 1994, near Raleigh, North Carolina, but the annual premium had already soared 23 percent, to about $80 million. The incentive for the airline, according to inside sources, would be the ability to recoup all or most of the additional premium should the NTSB rule American Eagle was not at fault in the crash of Flight 4184.

ATR hired a public-relations firm shortly after the negative press in November 1994. The politically affluent consortium was not afraid of wielding its clout, and even reportedly cajoled, threatened, and lobbied the news magazine show *48 Hours* up until several hours before airtime of its installment on airline crashes on March 2, 1995. The saga of ATR icing incidents was to have been a major portion of the *48 Hours* story. However, ATR's pressure seemed to have its desired effect, and the ATR segment was chopped to about 12 minutes of the hour-long show.

According to ATR's chairman and chief executive officer, Henri-Paul Puel, in an interview in 1990, the goal of the aircraft manufacturer was to sell 1000 of all ATR models before the turn of the century. By mid-1995, setbacks, competition, and a growing reputation for handling problems in ice had conspired to place ATR less than halfway to its goal after 10 years of operation. However, ATR's determination is not to be underestimated. A 1987 press report stated after the crash of an ATR-42 near Como, Italy, "the makers intended to defend the quality of their products and company image, by legal means if necessary, against grave and gratuitous judgments in the press on the reliability of the ATR-42."

Big money was at stake for the major players after the crash of Flight 4184. Although the FAA could apparently be viewed as bowing to pressure during earlier incidents, the NTSB was viewed as an independent agency. Although it is not unusual for conflicts and disagreements to be instigated over the findings of the NTSB, most aviation experts and even the media believe the NTSB to be above reproach.

The date for the final hearing on the crash of Flight 4184 was delayed from October to November 1995; then, hints of problems for the NTSB surfaced outside the agency. We learned from sources inside the government that the French were protesting the content of the report. In frustration, Cindy Gagliano called Greg Feith and asked what was going on. Greg confirmed the final report was in the third revision, and because of a protest, was now not scheduled to be released before January 1996. With tears in her eyes and voice breaking with emotion, Cindy Gagliano pleaded for information. Both Cindy and Al had neared the breaking point with the tension. The uncertainty of not knowing how the world would forever paint their son was damaging their family life and eating away at Al's health.

Feith agreed to discuss the issues within limits. He stated that the original report had been completed during July 1995 and submitted, except for the conclusions, to all parties for comment. The NTSB keeps the conclusions confidential until a public hearing. However, because of the International Civil Aviation Organization Treaty, the United States government was required to share the entire document with the French government. Rules prohibit the disclosure of contents of the report to interested parties, although exceptions are made. Our government shares the confidential details of foreign accidents involving U.S.-manufactured aircraft with organizations such as Boeing or McDonnell Douglas, and it would not be realistic to believe the French had not shared this one with ATR. The initial protest had arrived, reportedly on the 43rd day of the 45-day comment period, and the last one delayed the meeting another 60 days. Greg Feith shared the frustration of Cindy and Al Gagliano with the delays and sounded exasperated. He was, on that day, on his way to another meeting at the FAA and knew there would be pressure to modify some conclusions. "Things are changing almost daily," Greg told Cindy. "So there was no way to really determine what the final report would say."

In addition to the revising he had to do, Feith had to deal with political pressures from the FAA and ATR. Things were getting out of hand. Cindy was shocked to hear what he was telling her, but she was not too distraught to understand that what could be manipulation of the truth was going on. But she was not about to stop the flow of aggravated words from the investigator. "I feel like I have a bull's-eye on my back and on my butt, right now," Feith continued. "There's a lot of international networking and plenty of political things. I'm just as frustrated as you are." Cindy then added to his discomfort by reading him Gina Gagliano's letter to President Clinton begging for help and the response letter from the FAA. That information upset the investigator, and he requested that Cindy send him copies of the letters so he could use them in his battles with the FAA.

"What about the pilots, Greg?" Cindy pleaded, adding, "We need to know."

"Don't worry about the pilots," Feith responded. "I don't see anything with the pilots in the report as it stands now."

Of course, just because there was nothing to worry about at this point did not mean things would not change. Greg informed Cindy that pressure was coming not only from outside the NTSB but also internally as well. Even at the hearings in Indianapolis, he had been urged to back off by members of the technical panel, and his insistent questioning had landed him "on the carpet" more than once during the week. Before hanging up, Greg told Cindy that she, Al, or Gina could feel free to call him anytime with questions, even at home.

The answers provided Cindy with some consolation, but it gave everyone deep concerns over the outcome of the final report. If Greg could not withstand the pressure and downplayed certain facts to appease his superiors, the report could be tainted. At the least, it is reasonable to expect the final draft submitted to the full board for a vote is going to be less than it should be. The Air Line Pilots Association (ALPA) submitted comments on the investigation as the data existed in May 1995, as allowed under the rules of the NTSB. It also made some safety recommendations. Not surprisingly, the work was in-depth and could provide some insight into the final analysis of the technical staff, which will be submitted to the full board for a vote.

There is no question the crew was thoroughly trained, experienced, and qualified for the flight. According to the review committee of ALPA, the actions of the crew were "thorough, timely, and in complete compliance with company and Federal Aviation Regulation (FAR) requirements."

Others with a different perspective, such as ATR, might question the judgments and actions of the pilots for allowing the aircraft to continue holding for such an extended period in icing conditions. Some might consider the lighthearted conversations between pilots and flight attendants to be less than professional, and a few individuals might misunderstand the need for the captain to go to the bathroom while the flight circled. These are valid questions, and the answers given in the ALPA documents are valid appraisals of the situation. However, remember that ALPA has a stake in the outcome, just like the airline and ATR do. Although the conclusions of ALPA are similar to my own and those of highly respected aviation officials, it is doubtful that the final report issued by the NTSB will be as harsh on the aircraft and the FAA or as easy on the pilots and airline operational standards. The work contained in the ALPA comments was less than 60 pages and the NTSB document is more than 300, but both documents should be fairly consistent in the presentation of information.

The flight originated in Indianapolis and was operated under instrument flight rules on a filed flight plan as a scheduled Part 121 domestic passenger flight to Chicago O'Hare. The aircraft was flying in instrument meteorological conditions when the upset occurred, and the airplane was destroyed by impact. Evidence suggests an inflight breakup seconds before impact. All 68 persons aboard, 64 passengers and four crewmembers, were killed. The first officer was acting as the flying pilot, responsible for controlling the aircraft on this particular leg, and the captain was performing the duties of the nonflying pilot. A professional operational atmosphere existed in the cockpit, and all duties were conducted in a thorough manner.

An uncommanded deflection of the right aileron precipitated an automatic disconnection of the autopilot and caused the aircraft to roll violently to a right-wing-down position. The aileron deflection was a result of an unseen ridge of ice that developed behind the deicing boots on the wing's upper surface, causing a separated airflow over the aileron. When the airflow separated from the upper surface of the wing, a severe pressure differential developed between the upper and lower wing surfaces. As a result of extreme control forces resulting from the pressure differential, the flight crew was unable to physically overcome the airload forces, and the subsequent roll and dive were unrecoverable. The rate of aileron displacement was so rapid, human or autopilot input has been ruled out; neither the pilots nor the autopilot can be viewed as a causal factor in the upset.

According to information gathered subsequent to the accident, the ice ridge apparently formed in a position on the wing that was inconsistent with the pilots' previous experience and training and not defined in any documentation available to the crew. Because of the location of the suspected ice ridge at between 8 and 16 percent of chord and the high-wing configuration of the ATR-72 and ATR-42 aircraft, it is unlikely the pilots were able to see the suspect area of the wing. The crew was aware of ice accreting on the aircraft, based on statements contained on the cockpit voice recorder tapes, and all three levels of ice protection had been properly activated. No discussions of unusual ice development or patterns were recorded; no conversation took place regarding windshield or side window icing that would have indicated to the crew the presence of freezing precipitation. The crew appeared to have been comfortable with the operating environment and displayed no concerns. Conditions were obviously not deemed to be severe enough to warrant emergency or diversionary action by the crew, and the upset caught them by surprise. No indications have been found of ice-protection system malfunction, and the pilots had every reason to believe the ice was being properly removed from the aircraft. The progression of the flight is described in the following paragraphs.

The aircraft departed Indianapolis at 2:56 P.M. CST and proceeded normally along its route until making contact with the Boone Sector of the Chicago Center at 3:13 P.M.. The flight was issued holding instructions at the LUCIT intersection, located 18 nautical miles south of the Chicago Heights VOR, with an expect further clearance time of 3:30 P.M., which was updated to 3:45 P.M. The flight entered the hold at 3:24 P.M. and notified the company of the delay. At 3:38 P.M., the flight was advised of another 15-minute addition to the holding time. The captain commented about the high deck angle obtained during turns in holding pattern and suggested, "do you want to kick 'em in? [It'll] bring the nose down." After the first officer responded affirmatively, the flap handle was positioned in the 15-degree detent, and the deck angle, or angle of attack, was reduced to nearly 0 degrees. In addition to lowering the nose of the aircraft, the configuration change increased the airplane's airspeed safety margin.

At 3:41 P.M., the crew received a single chime alert that is believed to have been an audible indication the aircraft was accreting ice. Almost immediately after receiving the alert, the propellers were increased to 86 percent and the crew selected level 3 icing protection by activating the airframe boots. Seven minutes later, the first officer commented, "I'm showing some ice now." This was apparently the first visual indication by the flight crew that the aircraft was accreting ice. The first officer noted seven minutes later, "We still got ice," indicating he continued to have visual signs that the aircraft was either continuing the accretion of ice or had not completely shed the previous accumulation. The cockpit voice recorder suggested no unusual concern in the voice of the first officer.

At 3:56 P.M., Eagle Flight 4184 was issued a descent clearance to 8000 feet; however, because of the near-simultaneous audible traffic alert issued by the traffic collision and avoidance system, the crew failed to hear the instruction, and the controller issued it again 16 seconds later.

The flight crew placed both throttles in the flight idle position and began the descent. As the aircraft began its descent, the controller advised the flight "should be about 10 minutes until you're cleared in." This was the last transmission by air traffic control to Flight 4184. As the aircraft descended, the speed passed the maximum for flaps 15 degrees, and the crew received an aural warning, although the aircraft had not actually exceeded the 185-knot speed limit. The crew could have either slowed the aircraft to below the threshold speed for the warning or retracted the flaps; because of the imminent anticipated release from the hold, the crew elected to retract the flaps and placed the flap-selection handle in the fully retracted detent. The aircraft's pitch trim system activated while the flaps retracted toward 0 degrees and the "whooler" could be heard on the cockpit voice recorder tapes. This action was normal and consistent with the configuration change.

Suddenly, the aircraft's autopilot, which had been flying the airplane since shortly after departure from Indianapolis, disconnected, and an audible warning was picked up by the cockpit voice recorder. The flight data recorder also sensed a change in the discrete code for the autopilot status. Simultaneous with the autopilot disconnect, three thumps were heard on the cockpit voice recorder as a result of the control wheel impacting and bouncing on its full right stop. The crew was not aware of any previous controllability problems, and they were not concerned about the level of ice accumulated on the aircraft. A 24-second battle with the aircraft ensued as the pilots attempted to arrest the right roll and return the aircraft to level flight.

The digital flight data recorder indicated as the flaps approached the fully retracted position, the angle of attack increased to 4.7 degrees nose up. At that point, the right aileron deflected to a full right-wing-down, trailing-edge-up position. The aircraft, in response to the control surfaces, rolled violently to the right, to about 77 degrees right-wing-down. During the roll, the aircraft pitched down and its speed increased, decreasing the angle of attack. At about 2 degrees nose-up angle of attack, the flight crew regained partial roll control of the aircraft and the right roll rate was arrested. The aircraft rolled back to a 60-degree right-wing-down attitude, but the partial control lasted for only two to three seconds. The flight crew attempted to minimize the altitude lost and pulled back slightly on the control column to raise the nose, increasing the angle of attack once again, and at 4.7 degrees, the right aileron went full right-wing-down, and the aircraft began another violent roll to the right. Altitude continued to decrease as airspeed and vertical acceleration rapidly increased. Accurate flight data was lost at about 3500 feet above the ground with the aircraft traveling at more than 350 knots.

The cockpit voice recorder tape provided data that the upset took the crew completely by surprise. There were no prestall warnings or lateral control problem indications before the autopilot disconnect that could have alerted them of impending trouble. During the recovery efforts, cockpit conversations, although minimal, indicated the pilots also were attempting to confirm the reasons for the loss of control. The captain's statement of "autopilot's disengaged" indicated the captain had determined the upset was not the result of an autopilot malfunction. During the final 12 seconds, the captain's comments of "mellow it out" and "nice and easy" indicated the level of situational awareness of the aircraft's increasing speed, vertical acceleration, and decreasing altitude. These comments were a verbalization of the captain's concerns about overstressing or damaging the aircraft.

The aircraft hit the ground in a near-inverted, very steep nose-down attitude. Ground scars gave clear evidence of the inverted collision in that left engine components were recovered in the right crater and right engine components were discovered in the left crater. Evidence of an inflight breakup seconds before impact was gathered from the recovery of several aircraft control surfaces and composite wing skin surfaces along the flight path more than 1000 feet behind the point of main impact. The left outer wing section of about 10 feet was found behind and to the right of the flight path. The accident occurred at 3:57 P.M. CST. There were no preaccident communications with air traffic control indicating the presence of any flight crew concerns. Everything had proceeded normally, right up to the point of the loss of control.

The ALPA report considered the crew's appraisal of the icing conditions and whether proper decisions were made by the pilots before the upset in determining the operating conditions of the flight. It was obvious the experienced crew had no previous warning of the possibility of ice accretions behind the deicing boots and the potential threat to stability and control those accretions posed.

Additionally, a 1984 report by the Department of Atmospheric Science at the University of Wyoming indicated that encounters with large droplet conditions "were particularly insidious because the usual indications of icing (such as visible ice on the leading edges, windows, antennas, etc.) indicated that the accumulation was relatively minor." Although the critical ice accretion in the case of Flight 4184 seemed to have been located on the upper surface of the wing, this is primarily attributable to the location of the stagnation point with the flaps extended to 15 degrees at 170 knots indicated airspeed. Previous research aircraft encountered underwing icing in large droplet encounters mainly because the aircraft were either climbing or performing a slow-speed pass through the weather to maximize data collection, thereby exposing more of the lower wing to the ice.

The Rosemont ice detector probe located on the lower-left wing outboard of the engine on the ATR has proven effective in many types of icing environments, providing an aural cockpit indication to the pilots within seconds of entering the icing environment. It is improbable the pilots noticed ice before the probe sensed the accretion and warned the crew. As evidenced by their immediate action of selecting the final stages of ice protection, the pilots obviously and properly heeded the warning. Only two mentions of ice were made after the warning, and the use of the word "still" indicates the crew vigilance was proper.

The crew's decision-making process also had to consider the changing expect further clearance times that provided a framework for evaluating the operating conditions. The first remark from the crew was made just 12 minutes before the expect further clearance time; the second is just 3.5 minutes before the expect further clearance time and less than 30 seconds before the clearance for descent was issued by air traffic control. Because of the seven-minute elapsed time from the aural ice detection warning and the first visual cues of accretion obvious to the crew, the rate of accretion does not appear to have been particularly high. Crew discussions did not indicate the icing had changed or the rate increased between the verbal relay of observations. Although we will never know what the crew's response might have been to an additional delay from air traffic control, the knowledge of the impending expect further clearance time and issuance of an altitude change that could be attributed to the first stage of clearance out of the hold might have affected the crew's decision to not seek a routing or altitude change.

In the final analysis, the crew's actions can be viewed as proper within their experience, knowledge, and overall situational awareness. Had anything been unusual, more discussion would undoubtedly have occurred.

Numerous operational concerns came under intense scrutiny at the public hearings in Indianapolis, and the report by ALPA addressed some other issues that were investigated and publicized. The weather forecasts provided to the crews of Simmons Airline at the time of the accident included surface observations, terminal and area forecasts, and significant meteorological reports, or SIGMETs. Airline meteorological reports and Center weather advisories were available on request. No SIGMETs for severe icing were active on the night of the accident in the area in which Flight 4184 was operating, and freezing precipitation was not reported at any station between Indianapolis and Chicago O'Hare. Of the 11 pilot reports collected during the investigation that related icing encounters, three were for light rime icing; four were for light-to-moderate rime icing; two reported light mixed icing; one reported moderate mixed icing; one indicated light icing of an unreported type; and the final report indicated freezing rain. That final report is suspect, however, because the pilot reported negative ice accumulation, and any encounter with freezing rain would result in significant icing. The weather reports available to the crew, dispatchers, and controllers did not lead to the conclusion of any especially challenging icing encounters occurring during the day.

ATR criticized the crew for possible violation of company and FAA policies regarding the sterile cockpit, but significant questioning of parties at the public hearings failed to elicit an opinion from anyone that the flight crew had violated the rule. Several items dictate when sterile cockpit rules are to be enforced. First, the flight must be below 10,000 feet. Second, a critical phase of operation is occurring, such as takeoff, landing, and all ground movements. While holding at LUCIT intersection at 10,000 feet for more than 40 minutes, the aircraft was not operating in a critical phase of flight. Therefore, nonpertinent conversations, playing of the radio music, or leaving the cockpit to go to the bathroom are not considered forbidden, or even unusual, airline crew behavior.

The ATR uses a three-level icing protection protocol. Level 1 is always operated; level 2 is operated when encountering icing weather conditions; and level 3 is used when ice is actually accumulating on the airframe as evidenced by either visual cues or the Rosemont icing probe. Past experiences dictate that level 3 protection, once activated, must remain on as long as any ice is observed on the airframe. The cockpit voice recorder and digital flight data recorder information provided an indication of when the level 3 icing protection for Flight 4184 was activated.

Everyone involved in the investigation concluded the single chime aural warning was generated by the Rosemont icing probe, and the crew properly advanced the propeller revolutions from 77 to 86 percent and selected levels 2 and 3 of ice protection. Because activation of level 2 systems is not required until icing conditions are present, it is reasonable to assume the aircraft was operating between cloud layers in visual meteorological conditions previous to the warning with only level 1 ice protection systems operating. In this instance, the operation of the anti-icing and deicing systems was proper.

American Eagle and ATR procedures did not preclude the use of flaps 15 during holding at the time of the accident. At the heavy operating weight and with assumed intermittent turbulence, the crew's decision to place the flaps at 15 degrees not only resulted in a more comfortable ride, it increased the safety margin between minimum turning speed in ice and the operation of the flight at the maximum holding speed. The American Eagle-Simmons Airlines operating manual stated, "When holding is anticipated to be of a short duration, holding should be accomplished with the aircraft clean at a flap zero Conservative Maneuvering Speed. If a hold will be of an extended or indeterminate time period, the VmHB0 speed for Icing Conditions should be used as a holding speed." When Flight 4184 was placed in the hold at LUCIT, it was determined to be an extended hold. Therefore, per company manuals, they would have been required to fly at the speed of VmHB0 for icing conditions. This speed is defined as the "minimum speed for high bank operations with flaps retracted and is the minimum speed for the selection of HI-BANK mode on the AFCS." At the weight of nearly 44,500 pounds, the crew would have been using the 45,000-pound weight card and found the VmHB0 speed listed as 171 knots. The minimum speed for low-bank operations with flaps retracted was 165 knots. All turns in the hold were performed at high bank (25 degrees) with the exception of the final turn during the descent, which was selected at low bank (15 degrees). During the turns, while holding at 171 knots, the airspeed would drop to about 165 knots and the angle of attack would increase momentarily. Each time the airspeed dropped below 171 knots in a high-bank turn, the crew was losing its safety margin in icing conditions. By lowering the flaps to 15 degrees, the crew not only decreased the angle of attack, but increased the safety margin by 50 knots because the minimum speed in icing for high-bank operations with flaps at 15 was 121 knots. This was not only a prudent configuration change; it allowed for a safer flight and was how the crew had been trained to operate the aircraft. The crew was obviously comfortable in the environment, and incidental discussions with the flight attendants served to convey information the captain felt was appropriate for proper operation of the flight. On the basis of what the crew understood about hazardous ice formations and identifications, their actions and decisions are appropriate.

The *American Eagle-Simmons Airlines Aircraft Flight Manual—Part 1* defined the responsibilities and authority of the pilot in command. The pilot in command is charged with the supervision, training, and development in techniques, methods, and day-to-day flight activities, as well as discussion of crew activities with crewmembers, as necessary. Discussions with flight attendants under the captain's supervision were completely appropriate for the specific phase of Flight 4184; in addition, knowledge of the operating situation is required for crewmembers dealing with the general public. Performance of fuel calculations and operation of the addressing, communicating, and reporting system equipment for passing information of the status of the flight to company ground personnel, as well as obtaining gate data for the flight attendant's use, was well within the required operating duties of the flight crew.

The certification issues of the aircraft played a central role in the ALPA assessment of the investigative evidence and will most likely be strongly addressed by the NTSB as well. The certification rules for airframe ice-protection equipment are clearly defined in Part 25.1419 of the Code of Federal Regulations. No rules for determining adequate handling characteristics with ice accretion exist in the Code of Federal Regulations. Although the FAA does address some issues in an advisory circular, a multitude of techniques without criteria on acceptable handling and performance qualities exist from which the manufacturer can choose. Appendix C does not address freezing precipitation of any type. Therefore, no aircraft is evaluated for handling qualities with ice accreted from freezing precipitation of any type, and the functionality of ice-protection systems are not certificated for use in freezing precipitation. As a result, no data exists on the ability of any systems to properly protect an aircraft from such exposures. ATR did test the aircraft using Special Condition B6 under the Joint Airworthiness Rules, but the tests were confined to the Appendix C envelope. The *ATR Aircraft Flight Manual* warns, mainly because of the previous incidents, "Operation in freezing rain must be avoided." However, no information is provided on the identification or possible handling problems associated with an unintentional encounter.

FAA guidelines discuss the critical nature of large droplet precipitation and the need to consider the effects of ice accretion aft of the protected area of the wing. Because the pilot of an ATR cannot see much of the wing surface behind the boot, there is no way to determine whether the aircraft has been exposed to these conditions. FAA guidance also indicates an aircraft certificated for flight into known icing conditions must be able to withstand a 45-minute continuous exposure to icing conditions. If the aircraft cannot perform under the stated criteria, then the flight manual must state this. No such limitation is present in the *ATR Aircraft Flight Manual*. Gaps in the certification regulations exist. The debate over a proposed review and revision of the rules has raged for decades while safety organizations pleaded with the FAA to update the regulations to reflect the state of technology and understanding of the atmosphere. The debate goes on.

Just as holes exist in the certification regulations, gaps exist in the operational regulations as well. The rules for determining the extent of an icing hazard are highly subjective, and pilot understanding of mixed ice as a combination of rime and clear icing complicates the issue. If clear ice is a byproduct of freezing precipitation, any forecast of pilot report of mixed icing should be viewed as an outside of Appendix C encounter and therefore hazardous. During encounters with freezing rain or freezing drizzle, the hazard is the ice that accretes beyond the protected areas of the airfoils, and testing is usually accomplished for removal of ice on the boots. Testing with the Appendix C droplet sizes, which are smaller than freezing drizzle, is accomplished in the area between booted areas in the same longitudinal plane, not beyond. Normal ice accretion leads to large increases in drag, whereas the accretion of freezing drizzle or freezing rain may not. In the measurements accomplished on the data from the Flight 4184 digital flight data recorder, the presence of a reasonably inconsequential drag increase of 4 percent was noted, which led investigators to conclude the anomaly was not associated with any required pilot input, that is, the addition of power to maintain airspeed, which would have warned the pilots of a possible problem. The largest hazard posed by the ice accretion aft of the protected areas, as in freezing drizzle or freezing rain, may be flight control imbalances before a loss of lift on the wing. This may be attributable to the boundary layer disturbance originating farther aft of the stagnation point than the traditional icing-created wing stall. Behavior of the aircraft is anybody's guess because no engineering data can accurately predict the nearly infinite number of combinations of drop size distribution, temperatures, and lifting encountered in nature and transfer that data to workable theories on the effects on wing cross sections.

The ALPA report lambastes the FAA for inaction on important safety issues brought to its attention in 1982 and criticizes the agency's oversight and level of action in addressing past incidents with the aircraft.

Fourteen (14) years ago, the NTSB issued several Safety Recommendations to the FAA regarding aircraft certification as it relates to icing. Correspondence from the FAA to the NTSB contained numerous statements that were and are directly applicable to the current understanding of the industry in terms of aircraft icing certification. To date, these recommendations remain classified as either "Closed—Unacceptable Action" or "Open—Unacceptable Action." These recommendations, had they been acted on sufficiently and in a timely manner, would have prevented the accident of Flight 4184. The FAA must understand that Safety Recommendations issued by the NTSB are, in fact, in the true interest of safety, and should be given the attention they deserve.

The report cites a number of incidents and the responses from ATR that have already been discussed earlier. However, three previous events were selected, including the Como accident, and discussed in detail in the report to provide an accurate representation of previous incidents relative to the loss of Flight 4184 and the FAA inaction.

Two of the three above events took place in the United States, but the analysis was conducted by ATR or DGAC. The FAA must take definitive steps in studying trends in aircraft accidents and incidents, especially those that occur on U.S. soil. If foreign authorities conduct analyses on U.S.-registered aircraft, the FAA must take part in such a review. The FAA must verify and validate any rationale for modifications implemented on aircraft in the U.S.

The FAA was derelict in its duties to U.S. interests, ALPA asserts. No logical explanation exists for the release of Eagle Flight 4184 into an anticipated hold after a 40-minute delay on the ground. The system that places a turboprop airliner into a situation that forces it to be delayed more than 70 minutes with no effective means to track its progress in the system raises serious doubts about the efficiency of that system. ALPA recommended an overall review of the air traffic control system with emphasis on the management of air traffic. The flight from Indianapolis to Chicago O'Hare is typically 45 minutes in length. Eagle Flight 4184 was held 70 minutes in addition to its flight time. Sally Sprengel was the air traffic management coordinator on the day of the accident and stated that, when Flight 4184 was released, no flights were holding, but she had told them to expect the hold because of the large rush of traffic inbound from the west. ALPA concluded, "With an aircraft being delayed on the ground for over 30 minutes, and airborne holding being not only expected by Flow Control, but highly likely due to the possibility of saturated Sectors, the rationale for releasing an aircraft from a ground hold into a situation such as this is in need of review."

As an American Eagle ATR turboprop pilot operating out of Chicago O'Hare for more than five years, I can attest to a different level of air traffic control service for jets and turboprops. The report by ALPA provided detailed evidence to that aircraft-type variable approach to traffic handling, thus cementing the perception of pilots and passengers to another hazard of regional airline flying.

According to the air traffic controller's handbook, "operational priority" is defined as providing air traffic control services on a "first come, first served basis." However, Matthew Dunn, the traffic management coordinator for the Chicago area traffic control, stated in the public hearings that air traffic control will "do what you have to from an economic standpoint. . . ." A larger number of passengers connect from jets than do from turboprops. Turboprop aircraft bound for

Chicago O'Hare are handled by the Boone Sector and handed to approach control at less than 10,000 feet and descending to 8000 feet, whereas jets are taken directly from the high-altitude sectors with no speed restrictions. Under 10,000 feet, all aircraft are restricted to 250 knots, which means in the area near the airport, an ATR is operationally as capable as any jet. The current philosophy allows for jets to overtake and land before the turboprops. If the FAA is truly committed to a "one level of safety" campaign, as alleged by Transportation Secretary Federico Peña recently, then these issues of turboprops versus jets must be resolved.

Flight crews are trained to recognize the onset of a wing stall and apply standardized recovery techniques with an emphasis on a minimum loss of altitude. No stall alert was presented to the crew of Flight 4184, and recoveries from hinge moment reversals cannot be accomplished with minimal altitude loss. The pilots also were faced with a rapid onset of a violent, unusual attitude while operating in cloud. No unusual attitude recovery techniques were taught in airline training at Simmons Airlines at the time of the accident. Previous NTSB recommendations to the FAA included the provision that all flight training deemed hazardous be conducted in simulators only, where unusual attitude training can be accomplished, including recovery by reference to flight instruments only. There should be no illusion the crew of Flight 4184 would have been able to overcome the aerodynamic problems exhibited with the application of unusual attitude training; however, a future accident might be prevented by this causal side issue.

ATR pilots have long known of the side window trapezoidal ice formation as a sign of an encounter with freezing precipitation and, as a part of the airworthiness directive that released the ATRs back into known or forecast icing, had the requirement of specialized training to recognize the visual cues specific to the ATR. If pilots observe the trapezoidal ice formation on one or both of the side windows, they are required to immediately exit the area. At this time, the ATR-42 and ATR-72 are the only aircraft with such a visual cue and requirement. Any length of exposure must be considered harmful to the ATR aircraft, and an exact time period until criticality has not been determined.

The flight control system of the ATR is a manually powered, aerodynamically balanced system, meaning no hydraulic boost assistance is provided to position the control surfaces. During the initial design work for the ATR, a lateral control anomaly was documented that was believed to be similar in character to the phenomenon that caused the Roselawn accident. The original flight test data indicated that at angles of attack greater than 12 to 14 degrees, the force required to maintain a given aileron position (hinge moment) begins to decrease rapidly. For these reasons, a stall-protection system and vortex generators were installed on each model aircraft to provide suitable margin between normal aircraft operations and the anomaly. During the postaccident review of past incidents, an uncovered Aerospatiale document concluded that during a 1988 incident:

The ailerons tended to adopt the zero hinge moment position in the absence of pilot reaction. This deflection introduced a high roll rate. Two other roll excursions corresponding to increasing AOA [angle of attack] were checked by the control surfaces . . . probably due to the presence of ice on the airfoil beyond the de-icers, as is the case on all aircraft in freezing rain conditions.

Similar Aerospatiale analyses were made for incidents with Air Mauritius and Ryanair in 1991 and Continental Express in January 1994. Under the Bilateral Agreement, the DGAC was required to relay any applicable information to the United States. This information should have then been forwarded to the aircraft operators. Whether this was done is not known.

All turboprop aircraft were tested to see whether any other airliner of the type shared the problems and weaknesses exhibited by the ATR. Late in 1995, the results concluded that no other turboprop operated by U.S. airlines is as vulnerable to ice-induced flow disruptions as the ATR.

The ailerons on Flight 4184 deflected at a rate of about 52 degrees per second, meaning they reached the full deflection of 14 degrees in about one-third of a second. Such a rapid deflection is not possible with maximum autopilot rates of 9 degrees per second, and human pilots cannot exceed 30 degrees per second. This information alone indicated the existence of some outside force causing the deflection.

Within two seconds of the autopilot disconnection, Jeff Gagliano was wrestling with the controls. As the angle of attack decreased to less than 4.7 degrees, some amount of control was reestablished, and the bank angle lessened to 60 degrees. From that point on, the digital flight data recorder sensed aft control column pressure indicating a hands-on flight until impact.

As the vertical acceleration increased to a G-force of about 2.5, the trend in the aileron was to deflect further toward full right-wing-down. This could mean that control wheel force is a function of vertical acceleration, and the crew could not physically overcome the amount of control reversal force. ATR did testing on the phenomenon but limited the work to minimum vertical accelerations and airspeeds. The results of the testing have not been shared by ATR with other parties (such as ALPA) and the only released information concluded that the ATR meets FAA design limits. Significant testing in the realm of higher speeds and vertical accelerations is required to determine whether the ATR aircraft can be controlled by pilot strength alone.

The aircraft not only allowed the crew to enter an area outside the flight envelope with no possible entry back to controlled flight, but, in addition, the systems design of the aircraft interfered with pilot attempts at recovery. The multifunction computer has since been reprogrammed to allow ATR pilots to deploy flaps at greater than the self-limiting speed set in the computer by aircraft systems designers. The one option available to the pilots for recovering the craft had been prevented by a microchip.

Another limit on the pilots involves the travel-limiting unit on the rudder, which prevents crews from displacing the rudder more than 2 to 4 degrees either side of center when traveling at speeds greater than 185 knots. The mechanical stop is designed to prevent ham-footed pilots from overstressing the aircraft structure and inducing a fuselage-twisting moment by displacing full rudder at high speeds. Flight testing has indicated the need for extensive use of rudder to overcome the rolling forces encountered by aileron displacement, but this stop precluded the crew from anything more than minor, impotent deflections.

ALPA supplied the following conclusions to the steps leading to what it classified as a preventable accident. The steps began during the original design and certification of the ATR-42 and ATR-72 and still continue to this day.

The failure of the manufacturer to properly identify the cause of the aileron hinge moment reversal anomaly during design and certification and to adequately provide a systemic means to preclude such an event from occurring in flight.

The failure of the manufacturer and DGAC to properly notify the FAA of the aileron hinge moment reversal anomaly discovered during the original design and certification of the aircraft.

The failure of the FAA to take a more proactive role in the certification of aircraft manufactured outside of the United States.

The failure of the FAA to identify and correct potential dangers of an aileron hinge moment reversal exemplified through the demonstrated history of the ATR-42 and ATR-72 roll control incidents and accidents.

The failure of the manufacturer, DGAC, and FAA to adequately exchange information concerning findings and conclusions arrived at through the investigation of ATR-42 and ATR-72 roll control incidents and accidents.

The failure of the FAA to properly address repeated warnings issued by industry as to the existence of hazards when aircraft are exposed to "outside Appendix C" icing conditions.

As of late April 1996, the oft-delayed final hearing was scheduled for June 11, 1996. Continued wrangling with the manufacturer has repeatedly caused the NTSB to postpone this event, which would provide important closure for the victims' families. Asked again about the delays, investigator-in-charge Greg Feith told the Gaglianos that the 300-page report had been checked and rechecked numerous times and that he was ready for any objections yet to come from the French. He stated that the pressures he told Cindy about in October 1995 were still present and that the French government had "gone to the White House to force the issue." Greg has refused to bend to the pressure and said that he "will not change the report." As he told Jeff Gagliano's still-grieving family, "There is a big problem with this plane. Something has to be done."

17

Looking forward, looking back

A poster of the space shuttle *Challenger* blasting off from launch pad 39A at the Kennedy Space Center hangs on the wall behind my desk in my office at home. I have spent the last seven weeks in this room. Linde brought me meals that I have wolfed down to continue working on this story; sometimes I cry, sometimes I laugh, but I know its conclusion will bring a certain peace to many. I realized tonight with four days to go before my deadline that I have not really looked at the poster on the wall for a long time. Tonight, searching for the right words, I looked at the poster.

Brilliant, billowing clouds of snow-white steam tinged by bright yellow flame and deep orange reflections set against a deep blue, cloudless sky do not take away from majesty of the rising craft. The photograph captured an instant in time, an image of a machine that no longer exists. On another sunny day with a cold, bright-blue sky, *Challenger* and its seven occupants fell prey to the foibles of humans.

The stories are different. Ten times as many people perished in the crash of Eagle Flight 4184 than did in the *Challenger* tragedy. The January 28, 1986, explosion of the space shuttle, with its wildly twisting vapor clouds left behind by the solid rocket boosters, left graphic images in the minds of all Americans and will be forever remembered in history books and film clips. The newspaper articles containing vivid accounts of the destruction of an airliner falling to earth on a dreary October day in 1994 will lay yellowing in scrapbooks kept only by relatives and friends of those who perished.

The stories are the same. William Maready, the attorney who represented the widow of Mike Smith, the *Challenger* pilot, told me about the morning of the launch. The engineers were concerned about the degree of cold and ice on the space vehicle; there had been previous warning signs about the fragility of the O-ring seals in cold temperatures, and many opted for a delayed launch in the interest of safety. A telephone conference was held between Morton Thiokol, the builder of the solid rocket boosters, and NASA, wherein the concerns were discussed and the options weighed. A delay would cost millions of dollars, and those who were afraid of a serious launch risk were not successful in persuading others. Finally, the statement was reportedly made, "Gentlemen, it is time to take off your engineer's hats and put on your manager's hats."

The decision was made to proceed with the launch. One of the engineers calculated that a failure could occur between 60 and 65 seconds into the flight. He began breathing again as the clock passed the minute mark. Seventy-three seconds into the launch, it happened.

Pilots watched the evidence of an impending disaster with the ATR mount for years; we made statements after each incident to the effect that it was only a matter of time. Nobody thought it would happen to him. Then it did.

People looked the other way in both cases, ignoring a plethora of warning signals to keep the programs on track. The crew of seven on *Challenger* knew and understood the risks. The apparent success of the program had surprised everyone; there had not been a loss of vehicle and crew to that date. Statistically, it should have already occurred.

It was the same with the ATR aircraft. Unfortunately, nobody told the passengers or flight attendants of the risks they were being asked to assume. The pilots had been trained, knew the problems, and were extra cautious because of those warnings. Even so, they did not understand how close to the edge they ventured.

In the end, the die was cast when the decisions were made to launch. Rapidly, circumstances fell into the proper sequence for disaster, and there was no turning back. There is nothing, no power on earth, that could return us to a point in time before the disasters so that we could choose not to follow the same path. Most of us would give anything if it could be done. What matters is what we take from the experience—what was done and will continue to be accomplished in the aftermath of disaster.

I went to the Indianapolis Airport on Monday, October 23, 1995, to take photographs for this book. While I waited in the lounge for the arrival of an American Eagle ATR-72, I read the newspaper. On page A8 of the *Indianapolis Star*, I found a story from the Associated Press with a headline proclaiming, "FAA is still studying plane icing problems." The FAA had promised the complete retesting of all turboprop regional airliners in icing conditions to determine whether any of them shared the same vulnerability in icing conditions exhibited by the ATR. The agency had promised during a hearing in March that the testing would be completed by the anniversary date of the Eagle Flight 4184 accident, and that date was rapidly approaching. In the next few weeks, Drucella Anderson of the FAA said the tests would be complete and then decisions would be made whether to impose restrictions on the operations of the aircraft.

So far, Anderson stated, 18 other turboprops used in the United States had been investigated "at least once," but there were "two or three that the agency was going back to the manufacturers to request more information on." It would take several more weeks.

Reading further, I was amazed to find that, finally, the FAA was confirming what I had said nearly a year ago. Anthony Broderick, the FAA's assistant administrator for regulation and certification, stated, "The ATR's wing design is particularly sensitive to small disruptions [in the airflow]."

He was also quoted as saying something that really perplexed me: "The real issue is you don't want to hold in icing conditions. It's what has become a mantra of flying turboprops."

I almost said aloud, "But Mr. Broderick, holding is nothing more than flying around in a racetrack pattern. There is very little difference if you fly in a straight line for an hour at 200 knots to get from point A to point B or if you take your airliner and fly 30 minutes in a straight line followed by a 10-minute trek around a holding pattern and then another 30 minutes to your destination." Or is there?

I pondered the question for a few minutes before it struck me. Things are just the same as they were before the accident; things would always be the same with the FAA. The issue is not holding patterns but the vulnerability of the ATR aircraft to icing; however, by diverting the attention to something innocuous and not widely understood by the public, i.e., the holding pattern, the FAA shortcircuited a potential media frenzy about the airplane's performance deficiencies. The reporter quoted the FAA line without question.

The last two sentences of the story were also off the mark in relating the facts surrounding the Roselawn accident. "The FAA concluded the plane crashed because of a buildup of ice on the wings after it entered an unusual weather condition called 'supercooled drizzle drops'." Broderick should have known some of the originators of the supercooled drizzle drops, or SCDD, theory had already questioned the validity of the theory as it related to Eagle Flight 4184. He also should have realized it was not the responsibility of the FAA to determine the cause of the accident, even though its massive lobbying campaign with the NTSB was possibly beginning to have its desired effect.

"The National Transportation Safety Board, though, has delayed its formal finding of the crash's cause." It certainly has. I tore the article from the paper looked out the windows of the terminal. The FAA was apparently still susceptible to the powerful influences of a foreign manufacturer. We battled so hard against the system and won many battles; in the end, the FAA still decides who wins the war. That is a problem.

A friend of mine, an American Eagle ATR captain who had just been hired by a large jet carrier, called two days after my trip to Indianapolis. I was frustrated and concerned about the spin the FAA was putting on the news about the turboprops in ice, and it was good to hear a familiar voice of someone who really understood the problem and what was at stake. He said Ralph Richardi, the new president of AMR Eagle, had come to visit a winter operations training class held in Chicago for ATR pilots during late October 1995. Richardi was traveling around the Eagle system, my friend said, and wanted to talk about some changes that had been implemented at Eagle as they pertained to ATR operations. Some things were going to change for the better; some were going to remain the same.

A purge of the old regime had been underway at Simmons Airlines and American Eagle for the past several months. New management personnel who carried no ties to the old ways were brought on board. It seemed the outlook for the airline was fairly bright. The FAA and the Regional Airline Association agreed to set up an icing database to track the occurrences of incidents better so trends could be identified early. ATR winter operations training, which had been voluntary up to this year, was now mandatory for all crews, and the airline had set up a safety hotline for employees to report problems without fear of repercussions. I honestly was impressed and optimistic. Then came the downside.

My friend told me an ATR representative had been at Chicago O'Hare for the past week to answer any questions the pilots had about the weather or the performance of the aircraft. The ATR representative made statements in the winter operations class that the pilots attending the class believed to be blatant lies. "Nobody has problems with these airplanes except Simmons" was one of the falsehoods that infuriated the pilots in the meeting. The pointed questions the group asked of the factory representative were efficiently sidestepped, but Richardi took notes and promised to get to the bottom of the issues.

Recently when the weather turned cold and rainy, Simmons Airlines canceled 40 to 50 flights in one day. It could be that American Eagle was finally taking its future in its own hands and not hiding behind an FAA certification.

Ralph Richardi also related a story about jetways, the enclosed ramps that passengers use to board large jets, in use for the ATRs in Champaign, Illinois. He and his wife had recently traveled to the city to see the equipment in operation. Discussions of jetway use for AMR's biggest turboprop had been positive in corporate meetings. Even Robert Crandall, chairman of AMR Corporation, liked the idea. When the Richardis' flight from Chicago arrived in Champaign, someone had asked the flight attendants what the delay in disembarking was. "They are moving in the jetway," she responded, and a passenger standing next to the AMR president uttered, "Wow, big time!" That statement cut to the heart of the matter. As Richardi continued to relay the story to the pilots, he reportedly stated the reason Crandall wanted to investigate the possibility of using jetways at all Eagle airports: "With the jetway, people cannot see they are getting on an ATR," Richardi said. To that, an ATR captain replied, "Why don't we just put big shrouds around the propellers so people will think they are on a jet and not a turboprop?"

The real tragedy of American Eagle Flight 4184 is that many people saw the accident coming, but nobody acted definitively to prevent it. Everywhere I turned, I was shocked by what I interpreted to be a group of people maintaining a dangerous product on the marketplace because it was apparently cheaper to take the risk than it was to remedy the exposure. A respected journalist theorized during my research for this story that in her experience, "A conspiracy is nothing more than a bunch of widely separated individuals just trying to cover their own asses." That analysis fit, I thought. The bloated bureaucracy of the FAA did not lend itself to individual stands of conscience; the cluttered text of international treaties and concerns of international politics created a world more familiar to lawyers and deal brokers than the overworked employees of the government.

Knowledge, however, is power. If the flying public has been paying attention to the debate over the safety problems with the ATR, there will be an effect that cannot be denied by flowery explanations from itinerant bureaucrats. Chicago radio station WLS morning talk show host Don Wade joked about punishing convicted murderers by putting them on ATRs and flying them around in the ice, and the afternoon guys discussed the possibility of Bill Clinton being reelected: "His wings are all iced up, he's going down." The *Cleveland Plain Dealer* printed an editorial cartoon satirizing an FAA press conference wherein David Hinson announced the new safety equipment required by the agency for ATRs that fly in ice: a rabbit's foot, a four-leaf clover, and a lucky horseshoe.

The truth is the aircraft remains a hazard, as incidents of ice behind the new boot during summer operations have shown. American Eagle pilots who attended the winter operations class said that the class seemed evenly split between those who still feel there are problems and those who are convinced of the new safety of the ATR.

In the end, however, it is the decision of passengers to buy that ticket or board that aircraft that will determine the final fate of the ATR in the United States. I joke that I will not be happy until my diet soda can says on the side, "Packaging made from recycled ATRs." There is risk in everything, but it not unreasonable to assume some responsibility for your own decisions. Passengers do not have to walk across that ramp and get on an ATR when other options are available. Strapping yourself inside of an aluminum cocoon to be hurled through the air several miles above the earth is risky enough, but placing yourself inside 12 tons of aircraft that has no definable limits in icing is the realm of test pilots. Those folks get paid to push the operating envelope of the aircraft to see where the demons live; that experience should not be expected of a traveler purchasing a ticket on a scheduled airline.

On the human front there is good news. The Mission Un Impossible team of Helen Mudd, Chuck Mudd, and Julie Gunter successfully lobbied the Newton County Commissioners for permission to place a permanent monument at the crash site. In a flurry of activity, the memorial was planned and obtained; the bronze plaque was scheduled to be mounted on October 30, 1995. I cannot say enough good about these three selfless individuals and the others who have helped them.

Beginning the day they found Jenny Stansberry pulling weeds from around the roadside crosses placed by relatives, the Mission Un Impossible team has complied with every request. They mowed down weeds and cleaned up around the areas where families had placed tender reminders, they arranged for clergy to speak at an impromptu memorial, they constructed and painted 68 white crosses, and they lobbied a local flower shop for a matching number of potted flowers. They asked nothing in return and do not seek recognition. It took prodding to get them to allow me to tell their story in this book. Their efforts do not go unnoticed, and relatives and friends of the victims of Flight 4184 will not forget their names.

Recently, Tony Zarinnia showed me two plaques he designed. He wanted to give something to the families of the two pilots from a pilot. Tony's design is a brushed brass plaque inscribed with the pilot's name and an acknowledgement of his valiant efforts, framed on a background of blue velvet, under a duplicate of the Distinguished Flying Cross held by a red, white, and blue ribbon. At the bottom are pilot's wings supplied by American Eagle. Tony did not ask for recognition either; he did this to honor Jeffrey Gagliano, his friend, and Orlando Aguiar, a pilot he never met. They will be presented to the parents of Jeff and Orlando.

Stories and photographs from families have been trickling in to me all summer, but in the past two weeks, an avalanche of overnight delivery packages arrived at my door. Some days I cannot cope with opening another envelope to see pictures of loved ones spill across my desk. I have come to know many of the victims not only by name but also by pasting together details of their lives. Each story could fill a book, and I hope I have done them justice here. Although none of us will ever know them like their families did, they will not be forgotten. I cannot speak for others, but this experience has changed my life and the way I view our place in the hearts and minds of one another.

On a hot day in July 1995, I received a letter from the wife of one of the victims. She wrote of her husband of 33 years and summed up the feelings of so many in a succinct and poignant way:

"No amount of money will reimburse me when my daughter gets married and her Dad isn't there to give her away, or when I hold my first grandchild and Ken isn't there to share my joy, or when I'm alone in the night."

Epilogue

A year and a half has passed, and we still do not have the official final report on the cause of the American Eagle Flight 4184 accident. Government employees still struggle with the terminology to be used in the final draft of the report.

The Families of 4184—Roselawn organization fights on in spurts and has become a part of the national group of other air disaster survivors. Their fight is to smooth the road for those who will inevitably come down this painful path after us.

In 1995, another airliner fell from the sky after a propeller blade broke in flight. The pilots wrestled the aircraft to a crash landing that saved a number of passengers, but the accident cost others their lives, including the captain of the flight. The year 1995 did not meet the "zero accident" goal so loftily set by Transportation Secretary Federico Peña. Like the fabled pot of gold at the end of the rainbow, the goal is an illusion.

The Federal Aviation Administration owes it to the traveling public to execute its duty to the same degree of uncompromising professionalism as did the crew of Flight 4184. It must provide the United States with suitable, reliable, and safe aircraft to operate. We have had enough of the Band-Aid fixes and regulation by committee.

Part of the FAA's mandate is to promote aviation in the United States, and it takes that role very seriously. The FAA needs to put some of its overweight bureaucracy on the side of the scale dedicated to the regulation of aviation safety. Unfortunately, until a housecleaning is accomplished by dedicated professionals, the system will not change.

With the knowledge that the FAA might betray the flying public for the good of a foreign aircraft manufacturer, it is incumbent on all of us to make the best decisions for ourselves and our families. Even though your chances of dying in a Part 121 airline crash are about the same as getting hit by lightning on the golf course, you wouldn't tempt fate by running around outside with a 9-iron held high in the air during a thunderstorm. The next time you are about to board an

airplane and see a big propeller on it, pull an airline employee aside
and ask, "Is this an ATR?"

Accidents cost a great deal of money. Don't assume that I think
the airlines want to place undue risk on their operations. The airlines
must maintain public confidence in their services or they are out of
business. The competition is so fierce that even a perceptible hiccup
can be deadly. AMR Corporation lost millions of dollars in revenues
during the aftermath of the crash. Most likely these lost revenues will
be recovered from its insurance company or the aircraft manufacturer
after the dust settles; it can never recover from the images of the
spray of aircraft and humanity in that soybean field, but it will try.

ATR's goal of 1000 aircraft sold worldwide by the turn of the century will most likely never materialize, and even its advocates at the
FAA may realize the depth of ATR's problems—recent press articles
suggest that the FAA is beginning to distance itself from ATR. Time
will tell, but don't expect any miracles.

When Sonia Droy told me her story for the book, she said that
she had begun writing down her thoughts as she awoke from dreams
about John. One thing she stressed during our discussions was a note
she had scribbled: "Turn tragedy into triumph." I think we have attempted to do just that.

On October 31, 1995, Linde and I joined many family members in
Roselawn, Indiana. The weather was similar to that of a year ago,
with rain, icy mist, and temperatures that hovered around 40 degrees.
At 3:57 p.m., we stood in silence, holding lighted candles on the edge
of a farmer's field.

Bibliography

Air Line Pilots Association. August 28, 1989. Letter to FAA RE: In Qualified Support of 89-NM-101-AD.

Air Line Pilots Association. August 29, 1989. Letter to FAA RE: In Qualified Support of 89-NM-101-AD.

Air Line Pilots Association. May 15, 1995. Submission to the National Transportation Safety Board Regarding the Accident Involving Simmons (d.b.a. American Eagle) Flight 4184, report and cover letter.

Aircraft Technical Directory, IAP, Inc. 1980.

Airworthiness Directives 89-NM-101-AD, 89-NM-126-AD, T86-25-52, 86-NM-230-AD, 87-NM-165-AD, 90-NM-268-AD, 92-NM-126-AD, T-95-02-51, T94-25-51. FAA. 1986–1995.

American Airlines Sabre system. Passenger seat assignments, American Eagle Flight 4184.

American Eagle Flight Manual—Part 1. American Eagle. December 1994.

American Eagle Training Center. September 1, 1994. ATR-42/72 Ground and Flight Training Syllabus.

Anglemyer, Sandra, Plaintiff, v. AMR Corporation, AMR Leasing Corporation, Simmons Airlines, Inc., American Eagle, Inc., American Airlines, Inc., Avions de Transport Regional, and ATR Marketing, Inc., filed in United States District Court—Eastern District of North Carolina. December 27, 1994.

Articles pertaining to air traffic control issues. Various media sources.

Articles pertaining to aviation law and accident compensation.

Articles pertaining to bogus parts issues. Various media sources.

Articles pertaining to crash of American Eagle Flight 4184, ATR aircraft, commuter airlines, and family issues.

Articles pertaining to the FAA. Various media sources.

Articles pertaining to icing issues. Various media sources.

Articles pertaining to Midway Airlines and David Hinson. Various media sources.

Articles pertaining to wake turbulence issues. Various media sources.

Articles pertaining to windshear issues. Various media sources.

ATR Marketing. February 20, 1995. ATR Icing Test History, data.

ATR Marketing. October 31, 1994. ATR Key Figures, data.

ATR Support, Inc. January 6, 1995. Technical Background Paper.

ATR-42 Aircraft Operating Manual (AOM). ATR and American Eagle.

ATR-42 Flight Crew Operating Manual (FCOM). ATR.

ATR-42/72 Aircraft Operating Manual (AOM). ATR and American Eagle.

ATR-72 Aircraft Operating Manual (AOM). ATR and American Eagle.

ATR-72 Differences Manual. American Eagle.

ATR-72 Flight Crew Operating Manual (FCOM). ATR.

ATR. 1995. ATR Icing Conditions Procedures.

ATR. "ATR Icing Conditions Procedures, Version 1.0" brochure.

ATR. ATR-72 Briefing—January 1995.

ATR. November 5, 1995. "Freezing Drizzle—Towards A Better Knowledge And A Better Protection, Issue 1," brochure.

Attorneys for the Plaintiffs. 1995. "The Passengers and Crew—American Eagle Flight 4184—October 31, 1994," research document.

"Avions de Transport Regional ATR-42, Southern Morocco Accident," *Air Safety Week*. August 29, 1994.

Briot, Robert. February 28–March 2, 1994. "Icing Operations: Facing the Facts," technical paper, Aerospatiale Flight Test Directorate, 6th Annual European Corporate and Regional Aircraft Operators Safety Seminar, Amsterdam, Netherlands.

Broderick, Anthony. January 11, 1995. Memo for the FAA Administrator, briefing paper from the Associate Administrator for Regulation and Certification. FAA.

"Composites—The Lightweight Champs of Aircraft Industry," *Los Angeles Times*. March 9, 1987.

Dow, John P., Sr., and Gary D. Lium. February 1993. "Tailplane Stall: The Rime Is One Reason," technical paper. FAA.

FAA, DGAC and ATR. Documentation pertaining to the sale and registration of N401AM, an ATR-72.

FAA, DGAC, and ATR. ATR-42 Initial Certification Documents.

FAA, DGAC, and ATR. ATR-72 Initial Certification Documents.

FAA, NTSB, ATR, DGAC. 1986. Documents and reports pertaining to the June 6, 1986, tailplane loss of control incident.

FAA. October 30, 1987. ANM-100 Briefing Paper—Subject: ATR-42 Icing Incident.

FAA. December 1987. Review of ATR-42 Ice Protection, report and analysis.

FAA. July 29, 1988. Memorandum—Italian Request For Assistance RE: ATR-42 and associated correspondences and internal memos.

FAA. August 5, 1988. Memorandum from James T. Murphy RE: Italian ATR-42 accident of October 15, 1987 Inquiry Commission Hearing.

FAA. March 14, 1989. Letter to Air Line Pilots Association International RE: Information on ATR-42 icing certification.

FAA. September 30, 1992. Advisory Circular On Large Aircraft Ground Deicing.

Federal Register. Official Dockets for FAA Airworthiness Directives 92-NM-126-AD and 89-NM-51-AD.

FAA. June 23, 1993. Letter, Subject: ATR-42/72 Tailplane Stall Margin Analysis.

FAA, NTSB, ATR, Simmons Airlines and Crew Reports. Reports and documents pertaining to the tailplane icing incident in Marquette, Michigan, on November 24, 1993.

FAA. 1993–1994. Research Into Ice Contaminated Tail Stalls, memoranda and technical papers.

FAA. Air Traffic Control Documents pertaining to the handling of American Eagle Flight 4184, including FAA Form 8020-9, Aircraft Accident/Incident Preliminary Notice; FAA Form 8020-6, Report of Aircraft Accident; FAA Form 8020-6-1, Report of Aircraft Accident (Continuation Sheet); Flight Path Chart; Certified Indexes and Services Normal Statement; FAA Forms 7230-4, Daily Record of Facility Operations; Indianapolis Forms 7230-4, Indianapolis Air Traffic Control Tower Personnel Log IND/AGL; Chicago Air Route Traffic Control Center (ZAU) Form 7230-2, Personnel Log; FAA Form 7230-10, Position Log; Chicago Air Route Traffic Control Center (ARTCC) Automated Sign In Sign Out Program; Chicago ARTCC Flight Strips; Indianapolis Air Traffic Control Tower Transcriptions of Voice Recording; Chicago ARTCC Transcriptions of Voice Recording; FAA Form 8020-2.1, Facility Accident Notification Record; Personnel Statements; Weather Reports and Forecasts; Pilot Reports (PIREPS), Significant Meteorological Information (SIGMETS), Airman's Meteorological Information (AIRMETS).

FAA. Air Traffic Control Radar Data Plots for American Eagle Flight 4184.

FAA. December 5–16, 1994. National Aviation Safety Inspection Program—Inspection Report on Simmons Airlines, Inc.

FAA. December 29, 1994. 388 Certified and Uncertified Records of Service Difficulty Reports for the ATR-72 series aircraft, document listing.

FAA. December 29, 1994. 2,079 Certified and Uncertified Records of Service Difficulty Reports for the ATR-42 series aircraft, document listing.

FAA. December 30, 1994. Meeting Notes from Industry Meeting.

FAA. Federal Aviation Regulations.

Free Translation of President of Commission ATR-42 Accident—October 15, 1987. July 25, 1988.

George, Fred. April 1995. "Ice Protection And NLF Wings—Today's FAA certification envelope and the newer natural laminar flow wings won't guarantee your protection in all icing conditions," technical article. *Business & Commercial Aviation Magazine.*

Government Accounting Office. November 1992. "New Regulations for Deicing Aircraft Could Be Strengthened," report to the Ranking Minority Member, Subcommittee on Transportation and Related Agencies, Committee on Appropriations, U.S. Senate.

Hendricks, William R. (Director of Accident Investigation, FAA). May 28–29, 1992. "A Review of Aircraft Accidents Involving Icing/Deicing Issues," technical presentation, International Conference on Airplane Ground Deicing.

Hogan & Hartson. February 3, 1995. Letter to FAA regarding the demand of ATR to preview all documents released by the FAA under Freedom of Information Act, 5 U.S.C. 552, requests by news media outlets prior to their release.

Indiana State Emergency Management Agency. February 1995. "After Action Report."

Hopkins, George E. 1982. *Flying The Line—The First Half Century of the Air Line Pilot's Association.* Air Line Pilots Association.

"Icing Trial on the ATR-42 in A&AEE Blower Tunnel," report, Aeroplane and Armament Experimental Establishment—Bascombe-Downs, England. April 1988.

Indiana State Police Report, filed November 11, 1994.

"Inflight Icing Parts I, II, and III," article and study report. *Air Line Pilot Magazine.* August, September, and October 1995.

Ingelman-Sundberg, Martin. January 1992. "Why Icing Causes Tailplane Stalls," technical paper.

International Federation of Air Line Pilots Associations. November 30, 1989. Safety of the ATR-42 in Icing Conditions, memo for worldwide distribution.

Jane's Encyclopedia of Aviation. 1989.

Kost, Garrison, and James B. Clapp. "Reconstructing An Accident: Expert Testimony Building Blocks."

Leonard, Glenn (AMR American Eagle Fleet Manager). 1994. "Some Significant Dates/Events in ATR Evolution," technical paper. American Eagle.

Lincoln Township Volunteer Fire Department logs and reports.

National Transportation Safety Board and Federal Aviation Administration. 1981–1994. Correspondence and recommendations.

National Transportation Safety Board. September 9, 1981. "Aircraft Icing Avoidance and Protection," report.

NTSB, Simmons Airlines, Crew Reports, FAA, and ATR. Documents and correspondences pertaining to the two inflight loss of control incidents occurring at Detroit, Michigan, December 18, 1986.

NTSB, Simmons Airlines, Crew Reports, FAA, and ATR. Documents and correspondences pertaining to the inflight loss of control incident occurring at Central Wisconsin Airport, December 22, 1988.

NTSB. March 4, 1993. Factual Report RE: Britt Airways/Continental Express Flight 3444—ATR-42 Inflight Icing Incident, report summary.

NTSB. November 18, 1994. List of Incidents, ATR related incidents listing.

NTSB. October 31, 1994. "National Transportation Safety Board Preliminary Report—American Eagle Flight 4184."

NTSB. December 13, 1994. "Joint Flight Standards Information Bulletin for Airworthiness and Air Transport," document. NTSB recommendations: A-94-182, A-94-183, A-94-184, and A-94-185.

National Transportation Safety Board investigatory exhibits, February 27, 1995.

National Transportation Safety Board witness testimony transcripts, audio tapes, and videotapes from public hearings in Indianapolis, Indiana, February 27–March 3, 1995.

"Partial CVR Transcript—Royal Air Maroc ATR-42 Crash," transcript. Undated.

Perkins, Porter J. 1994. "Tailplane Stall Caused By Ice," technical paper, 39th Corporate Aviation Safety Seminar—Flight Safety Foundation, April 13–15, 1994.

Perkins, Porter J. January 11–14, 1993. "Aircraft Icing Problems—After 50 Years," technical paper, 31st Aerospace Sciences Meeting and Exhibit, Reno, Nevada.

Roskam, Jan. December 1994. "Ice Can Be Deadly On Your Tail," technical paper. *Professional Pilot.*

Shaw, Adam. 1977. *Sound of Impact.* Viking Press.

Simmons Airlines. August 1989. Original 1989 ATR Systems Training Syllabus.

Simmons Airlines. Issued August 1989. *Simmons Airlines General Operations Manual.*

Simmons Airlines. September 1, 1991. *American Eagle Aircraft Operating Manual—Volume 3.*

Simmons Airlines. 1992. "Simmons Airlines Winter Operations Handout—1992."

Simmons Airlines. November 4, 1992. *American Eagle Aircraft Operating Manual—Volume 2.*

Simmons Airlines. 1993. "Simmons Airlines Winter Operations Handout—1993."

Simmons Airlines. October 31, 1994–June 20, 1995. Various electronic mail messages.

"Simmons Airlines Flight Operations," newsletter. December 1993.

St. Laurent, Bob. March 5, 1995. "Toward Safer Airlines," research paper.

Steenblick, Jan W. January 1992. "Turboprop Tailplane Icing," technical paper.

Steenblick, Jan W. August 1995. "Inflight Icing: Certification vs. Reality," research article. *Air Line Pilot Magazine.*

"Straight and Level," Simmons Airlines ALPA Master Executive Council (MEC) publication. December 1994/January 1995, February/March 1995, April/May 1995, June/July 1995, August/September 1995, October/November 1995.

System Safety, Inc. December 10, 1987. Preliminary Analysis report to Alitalia Airlines, RE: October 15, 1987 accident of ATI Flight 460.

U.S. Department of Transportation. 1975. *Aviation Weather.*

U.S. Department of Transportation and the Federal Aviation Administration. June 1983. "A New Characterization of Supercooled Clouds Below 10,000 Feet AGL," technical report.

Witness statements of Clarence Hanley, Lynn Marie Stone, Robert D. Hilton, Larry Midkiff, Carol Prohosky, Norman Prohosky, Brenda Lee Quisenberry, Charles Wann, and Robert Stone, recorded by Allen Ryan.

Index

Illustrations are indicated by **boldface**.